Japanese Lesson Study in
MATHEMATICS

Its Impact, Diversity and Potential
for Educational Improvement

Japanese Lesson Study in
MATHEMATICS

Its Impact, Diversity and Potential for Educational Improvement

Editors

Masami Isoda
University of Tsukuba, Japan

Max Stephens
University of Melbourne, Australia

Yutaka Ohara
Naruto University of Education, Japan

Takeshi Miyakawa
University of Tsukuba, Japan

With support of

Shizumi Shimizu
Takuya Baba
Kazuyoshi Okubo
Abraham Arcavi

World Scientific

NEW JERSEY • LONDON • SINGAPORE • BEIJING • SHANGHAI • HONG KONG • TAIPEI • CHENNAI

Published by

World Scientific Publishing Co. Pte. Ltd.

5 Toh Tuck Link, Singapore 596224

USA office: 27 Warren Street, Suite 401-402, Hackensack, NJ 07601

UK office: 57 Shelton Street, Covent Garden, London WC2H 9HE

British Library Cataloguing-in-Publication Data
A catalogue record for this book is available from the British Library.

JAPANESE LESSON STUDY IN MATHEMATICS
Its Impact, Diversity and Potential for Educational Improvement

ISBN-13 978-981-270-453-5
ISBN-10 981-270-453-1
ISBN-13 978-981-270-544-0 (pbk)
ISBN-10 981-270-544-9 (pbk)

Printed in Singapore by World Scientific Printers (S) Pte Ltd

Contents

Chapter 1: Japanese Lesson Study in Mathematics

Section 3: Official In-Service Teacher Training System

Section 4: Mathematics Curriculum and Way of Implementation

Section 5: Comparisons of Features of Past International Comparative Studies

Chapter 2: Methods and Types of Study Lessons

Chapter 3: Trends of Research Topics in Japan Society of Mathematical Education

Section 1: Lesson Study in Elementary Schools

Section 2: Lesson Study in Junior High Schools

Section 3: Lesson Study in High Schools

Chapter 4: Diversity and Variety of Lesson Study

Chapter 5: International Cooperative Projects

Appendices

Foreword

Japanese Lesson Study in Mathematics: its impact, diversity and potential for educational improvement has been prepared for non-Japanese Mathematics Educators who are interested in learning about the development and use of Lesson Study in Japanese Schools and its growing use in other countries as a means of improving teaching and learning in mathematics.

This book has been translated from a Japanese version, *Zudemiru Nihonno Sansu Sugaku Jyugyo Kenkyu* edited by Shizumi Shimizu, Masami Isoda, Kazuyoshi Okubo and Takuya Baba, and published in 2005 by *Meijitosyo* publishers, Tokyo.

One important difference is the inclusion in this book of four commentaries on sample Japanese lessons which have been prepared by Abraham Arcavi and Aida Yap. The lessons on which these commentaries are based can be downloaded from the following web site http://www.criced.tsukuba.ac.jp/math/video/. These commentaries are intended to help non-Japanese readers to understand the structure of the above lessons. Particular attention is given to describing the mathematical contents of the lessons, their main components and key events in the classroom. There are also possible issues for discussion and reflections with teachers observing these lessons.

This book explains the historical development of Lesson Study in Japan and relationships between Lesson Study and the National Course of Study. The various authors explain how Lesson Study activities are supported by all elements of the Japanese education system. Samples of lessons developed for Lesson Study are included. These give practical examples of the careful planning of content, teaching approaches and expected students' responses that are an essential part of Lesson Study. Also included in the book are several case studies showing how Lesson Study has been used in a number of overseas locations.

Much more could be said about the development and use of Lesson Study than has been included in this book. However, we hope that it will help introduce overseas readers to this important practice in which teachers and researchers work together to improve students' mathematics.

We would like to acknowledge with contribution of Abraham Arcavi for his developing of original appendices in English. Many staff members of the Center for Research on International Cooperation in Educational Development (CRICED) at the University of Tsukuba have assisted the editors of this translation to prepare final copy for publication. In particular, it is necessary to acknowledge with appreciation the work of Arturo Mena-Lorca, Rie Sugiura, Takafusa Okamura, Kimiho Chino, Kazuhiro Aoyama, Atsuko Ishikawa, Kyoko Obayashi and Hidetomo Machi for their care and assistance in preparing this translation for printing.

The key agencies that have been responsible for making this book possible have been acknowledged in the *Introduction* to the original Japanese version.

Editors   Masami Isoda

Max Stephens

Yutaka Ohara

Takeshi Miyakawa

Introduction: to the English Translation

This book on the Japanese Lesson Study is a compilation of contributions from various scholars in mathematics education in Japan. The original text was prepared in Japanese. This translation, therefore, is intended to give those interested in this important feature of education in Japan an overview of how the practice of Lesson Study evolved from the late nineteenth century and its continuing relevance today. For some non-Japanese readers, the idea of having other teachers come into one's classroom and observe one's teaching, and later to discuss, and possibly critique, what they observed may be quite strange. Teachers in all countries, of course, discuss their lessons with their colleagues. Around any staff room it is possible to hear comments such as: "Today's lesson went really well", or "I struggled today to get across the idea of ....", or "How do you introduce students to negative numbers?". But entering another teacher's classroom, except to give a brief message, may still be seen as intruding on a colleague's personal space. Indeed, it is possible for some teachers never to observe another teacher's lesson after graduating from teachers' college. How are things so different in Japan?

The notion of a teacher's classroom space as private professional space has never been strong in Japan. In other countries, this notion appears to be disappearing with the practice of several teachers sharing the same classroom through team teaching. In training sessions for programs such as *Reading Recovery* in the elementary school, the use of "behind the screen" observation of teaching has been shown to be a powerful tool for having a group of teachers observe a lesson (from behind a one-way screen so as not to be seen to intrude on children's learning and a teacher's teaching). A similar model has been used in Australia in the professional development of teachers of mathematics in the elementary school. Together with an expert trainer, those observing the lesson usually have with them the class teacher's plan of the actual lesson, and are asked to comment on what they are observing and to relate what is happening in the class to the teacher's lesson plan.   Later the class teacher will be invited to join the group for discussion and further reflection. In some cases, lessons are both observed and videotaped for subsequent training purposes.

Key ideas underlying Lesson Study in Japan

In many ways, these practices reflect the key ideas of the Japanese Lesson Study.   As Professor Hattori remarks later in this book, "Lesson Study does not refer just to in-school training (or, in our words, simply to observing another teacher's lesson). It is a process by which teachers of mathematics at several schools in the same community work together to research teaching materials, develop teaching plans (lesson plans) and practice teaching lessons."

Underlying the practice of Lesson Study is the idea that teachers can best learn from and improve their practice by a seeing other teachers teach. Second, there is an expectation that teachers who have developed deep understanding of and skill in subject matter pedagogy should be encouraged to share their knowledge and experience with colleagues. Thirdly, while the focus appears to on the teacher, the final focus is on the cultivation of students' interest and on the quality of their learning. The various cycles of refinement which are at the heart of the Japanese Lesson Study only make sense in terms of improving the quality of student's learning.

In ***Before It's Too Late****: A Report to the Nation from the National Commission on Mathematics and Science Teaching for the 21ˢᵗ*

Century (2000) the authors (p. 20) quote from James Stigler's conclusions from various videotape research studies of mathematics teaching that "The key to long-term improvement [in teaching] is to figure out how to generate, accumulate, and share professional knowledge." The Japanese Lesson Study has proved to be one such means. The authors of *Before It's Too Late* commend the widespread use in Japan of problem-focussed lessons involving collaborative work by students carefully guided by the class teacher. They refer to this approach to mathematics teaching as "a natural outgrowth of the culture of teaching and school life in Japan where teachers are accorded not only abundant time for lesson preparation, but also for collaborative lesson planning" (p. 20).

Lesson Study is a key means of supporting and fostering collaborative lesson planning throughout the compulsory years of schooling in Japan. It also features in the senior high school. The authors of **Before It's Too Late** claim that almost all teachers of elementary schools and about half of all middle school teachers participate in Lesson Study groups that meet for two to five hours each week (p. 20). This statistic is clearly intended by the authors to dispel any sense of complacency among readers in U.S.A. While two hours per week may well be the more typical, this time is certainly made up from some official meeting time, but also with after-school meetings and meetings held in teachers' own time. How is this practice among Japan's teachers and schools so different from the experience of teachers in many other countries?

How Lesson Study is supported and promoted in Japan

Such a major commitment of time and resources is possible only when many elements work together to give a powerful endorsement to the practice of Lesson Study. At every level of the school system, Lesson Study is strongly promoted. In the first place, all teachers are seen to have an obligation to improve their lessons and to engage in continuing study. For many teachers, therefore, participation in Lesson Study related activities in their own school represents a clear and rewarding way of meeting this expectation.

Support from within the school system

School principals support teachers' involvement in Lesson Study by conducting and setting Lesson Studies in the course of the school year. Principals are also in a position to endorse teachers' contributions to Lesson Studies outside school time. Teachers' societies which are managed by teachers and principals also foster Lesson Studies. In each school district, the board of education also organizes ongoing programs relating to Lesson Study which are conducted in working hours.

Some principals conduct an open forum every two or three years to which teachers from other schools are invited to visit classrooms and to see teaching which has been cultivated through Lesson Study in the host school. In these open forums, visitors are able to see the results of sustained application of Lesson Study for both teachers and students – in particular, how well teacher has taught and how well a teacher has cultivated student learning. In this sense, the quality of what students have learned over time is showcased for all to see. The fruits of good teaching are expected to be observed directly. This is a helpful corrective to those outside Japan who believe that student learning can only be inferred from tests scores and other external performance data.

Voluntary commitment from teachers

In addition to these clear official endorsements of Lesson Study, a strong sense of voluntary commitment underpins teachers' participation in Lesson Study. Teachers see direct rewards for their efforts when their students have learned mathematics well. There is strong sense of collegial learning and self-actualization when teachers choose to work with and learn from other teachers. Teachers can readily see evidence of their own professional growth and at the same time come to appreciate the beneficial effects of Lesson Study on their students' development. Some teachers are also committed to enhancing specific reforms, especially through teachers' independent associations.

After formal Lesson Study activities in their school or in a host school, it is quite common in Japan for teachers to conclude the day's events with an *enkai* or small evening banquet with drinks. During these more relaxed parties, participants are often invited to express freely their ideas and reactions to what they have observed during the day. This kind of social event acts as an important and productive way of closing the day's activities.

Publishers

There is also a very clear sense in which the commercial publishing system values and promotes the products of Lesson Studies. Major publishers in Japan publish teachers' lessons that have been developed through Lesson Study. Talented and experienced teachers have a chance to publish their best practice and can feel a sense of professional recognition and active involvement in various reform movements. Publishers also support Lesson Study by publishing resource materials and ideas for lessons that have been written by teachers themselves.

Universities and supervisors

There is a direct and clear involvement of universities and other elements of the education system in Lesson Studies. University researchers and supervisors are requested to join teachers in and to contribute their knowledge and experience to Lesson Study activities. In collaborating with university researchers, teachers' contributions are recognised not so much through the publication of academic papers, but primarily through their quality of their classroom practice and the evident achievements of their students.

Supervisors have a special role to support teachers' practice based on policies and advice emerging from the Ministry of Education. In Lesson Study activities, supervisors play a pivotal role in making constructive evaluations to individual teachers whose lessons they have observed and in appraising and synthesizing comments made by teachers in ensuing discussions, drawing out and highlighting key points. To each of these tasks, supervisors are expected to bring a profound knowledge of the National Course of Study and their own wide experience of participating in and of leading Lesson Studies in their earlier careers.

University researchers are also expected to bring to Lesson Study a deep knowledge of the National Course of Study. While they may not have direct or extensive experience of classroom teaching, University researchers are expected to have accumulated deep knowledge of teaching practice through observation and participation in Lesson Study activities so that they can provide constructive and well informed comments on lessons observed and the ensuing discussions. University researchers may also be invited to join specific Lesson Study activities based on their known research interests and expertise.

A career-long involvement in Lesson Study

It is clear that participation in Lesson Study touches and involves teachers at all stages of their professional lives. Lesson Study has a direct bearing on a teacher's own career advancement since promotion is determined by the principal in concert with the local board of education. Beginning teachers are required to participate in Lesson Studies. Participation in Lesson Study activities will be an important element of their early professional development, and will possibly be the most important formative influence on their teaching. Experienced teachers are expected to lead Lesson Study activities in their own school, and to contribute to Lesson Studies held outside their home school. Being recognised as a leading teacher through Lesson Studies is a mark of high professional recognition in Japan.

For those who move on to become sub-principals or supervisors there is a clear responsibility to organise Lesson Study activities through school-based programs or through other programs at local level. Overall responsibility for conducting and planning Lesson Study through a school year is a key role for school principals acting together with the local board of education.

About this book

This book is divided into five chapters.

The first chapter gives a brief definition and history of the development of the Lesson Study in Japan. It also describes the different contexts in which Lesson Study operates – some formal and mandated, some quite informal and voluntary. There is also a discussion of how official policy documents have introduced new themes or priorities for mathematics teaching and how these in turn influence topics for Lesson Study at local level. This chapter concludes with an important discussion on formative assessment and its relation to Lesson Study.

The second chapter provides some detailed examples of lesson plans and study lessons. These plans have a very clear focus on promoting opportunities where students can take the initiative in their own learning.   For this to happen the teacher has to create situations which offer students different ways to develop their own thinking. The teacher has to pay close attention to how students are thinking and through questions help them to refine their thinking. Several different models of lessons are presented: problem solving oriented teaching, discussion oriented teaching and problem discovery oriented teaching, and these models are applied to several different and important mathematical topics for the elementary school.

The third chapter deals with research trends in annual meetings of the Japan Society for Mathematical Education (JSME).   Research topics are discussed from the perspective of the elementary school, the middle school, and the senior high school.   There are important differences in emphases among these sections.   In the case of the elementary school, the focus is on general ideas in education.   These include training in basic ability, learning how to learn, enjoying mathematics, appreciation of mathematics and thinking mathematically.   Individual learning and learning in small groups are also considered.

For the middle school, the focus shifts more clearly to particular topics for mathematics learning and to methods of teaching. Particular reference is made to problem-solving and project-type learning.   Attention is also given to students' more advanced mathematical thinking, such as relational and functional thinking, and to their use of technology.

High school topics have a very clear focus on particular courses of study, individual topics and associated teaching approaches.

Chapter four illustrates the various supporting relations that sustain Lesson Study.   Some of the relations considered in this chapter are those between a university and Lesson Study activities conducted in its attached (demonstration) school. (In the case of national universities, a professor from the university occupies the formal position of school principal of an attached school). Then there is the partnership between universities, boards of education and local schools. For example, training programs provided for teachers during working time frequently include Lesson Study, especially in the case of first-year teachers who will present their own study lessons. Lesson Study clubs (sometimes called Lesson Study groups or Lesson Study committees) are less formal structures centred around Lesson Study activities by teachers acting on their own initiative, usually in their free time, and sometimes involving teachers from across several schools.

The final chapter deals with international aspects of Lesson Study

In the first section of this chapter, several international research projects involving a comparative study of lessons from several perspectives are outlined. Other parts of this chapter focus specifically on attempts to utilise the principles and practice of the Japanese Lesson Study in a number of overseas countries through the efforts of Japanese trained mathematics educators. Many of these missions have been supported by the Japan International Cooperation Agency (JICA). These case studies show the resilience of the Lesson Study when taken outside its home country, and its capacity to act as a powerful stimulus for the improvement of teaching in other countries. Moving outside its own historical and cultural context may entail the loss of some of the powerful influences that shape and give direction to Lesson Study in Japan. But on a more optimistic note, we should note from these case studies that Lesson Study when used in other countries can keep a strong sense of identity and purpose that have made it so effective in Japan. We should also expect that it might itself undergo creative transformations as a result of being grafted onto different cultures.

About further readings

This book omitted huge references because all most references are written in Japanese. For non-Japanese readers, it is impossible to find and read Japanese books. In page 28, there are announcements for further readings in English published from Japanese Society of Mathematics Education. In the site URL: http://www.jica.go.jp/english/resources/publications/study/topical/educational/, there are very good resources which introduce Japanese Education in English, Spanish and French.

Joint Editors of the English translation

Max Stephens
The University of Melbourne
Australia

Masami Isoda
University of Tsukuba
Japan

References

National Commission on Mathematics and Science Teaching for the 21ˢᵗ Century (2000). **Before It's Too Late**: *A Report to the Nation from the National Commission on Mathematics and Science Teaching for the 21ˢᵗ Century. Washington, D.C.: author.*

Introduction: Translated from the Japanese Version

In December 2004, the results of the Programme for International Student Assessment (PISA) and Third International Mathematics and Science Study (TIMSS) were announced by the Organization for Economic Co-operation and Development (OECD) and International Association for the Evaluation of Educational Achievement (IEA), once again drawing attention to the need to foster and further improve students' mathematics performance. In Japan this is going to require further Lesson Study and improvements to lessons based on the results of Lesson Study.

This book shows how Lesson Study is to be conducted to improve classroom teaching. It has been compiled in the hope of improving the quality of education by showing how successful teaching methods have thus far been developed by passionate teachers and research organizations in light of curricular revisions and changing trends, and by demonstrating how this experience has contributed to improving education worldwide.

One of the triggers that drew worldwide attention to Japanese teaching methods was a comparative study on problem solving between Japan and the USA that began in the 1980s (USA representative Jerry Becker, Japanese representative, Tatsuro Miwa). The high regard for mathematics classes taught in Japan became widely known among educators in the 1990s thanks to research conducted by James Stigler and later published by Stigler and Heibert (1999) in "The Teaching Gap: Best Ideas from the World's Teachers for Improving Education" (New York: Free Press).

"Lesson Study" (jugyou-kenkyu) thus came to be known around the world as a uniquely Japanese method of lesson improvement designed to facilitate the development of high quality lessons, and resulted in a Lesson Study boom in the industrialized nations, particularly the USA. In the meantime, our cooperative efforts in education in the developing countries yielded the 1990 acceptance of the "World Declaration on Education for All," and proposals were made to address issues like "improving the quality of education" and promoting "excellence" especially on "numeracy" – mathematical literacy for learning, the ability such as to perform calculations and deduce mathematical reasoning useful in everyday life, mathematical intellect – (see "World Education Forum", Darker 2000). Since then, improvements in mathematics education have been counted among the core components of international cooperation in education, and today Japan's problem-solving oriented teaching methods and Lesson Study techniques have come to serve as a useful reference tool around the world.

As a joint plan by the Research Department of the Japan Society of Mathematical Education and Cooperation Bases System Mathematics Committee for International Cooperation in Education under the Ministry of Education, Culture, Sports, Science and Technology, this book is intended to serve as a handbook for teachers engaging in Lesson Study, as a textbook that can be used for teacher development courses and teacher training programs, and as basic material for informing overseas observers about Japanese Lesson Study practices.

Special thanks are due to the Ministry of Education, Culture, Sports, Science and Technology (MEXT): Office for International Cooperation, International Affairs Division, Minister's Secretariat, and to Japan International Cooperation Agency (JICA) for supporting this publication.

We sincerely hope that this book will be able to provide observers both inside and outside of Japan with basic information that can be used to promote improvements to teaching practices.

March 2005

Editors of Japanese version

Shizumi Shimizu
Director
Research Department
Japan Society of
Mathematical Education

Masami Isoda
Representative
Mathematics Task
Cooperation Bases System
by MEXT

Kazuyoshi Okubo
Secretariat
Mathematics Task
Cooperation Bases System
by MEXT

Takuya Baba
Secretariat
Mathematics Task
Cooperation Bases System
by MEXT

CHAPTER 1

Japanese Lesson Study in Mathematics

Japanese Education and Lesson Study: An Overview

Section 1.1: "How is Lesson Study Implemented?"

Takuya Baba

Introduction: Lesson Study, currently a topic of worldwide attention, refers to a process in which teachers progressively strive to improve their teaching methods by working with other teachers to examine and critique one another's teaching techniques. First developed as an educational practice in the Meiji period of Japan, Lesson Study functions as a means of enabling teachers to develop and study their own teaching practices. It is this function to which its international attention can be attributed.

1. The Process of Lesson Study

Lesson Study consists of *preparation, actual class, and class review sessions* in Japanese "kyozai kenkyu", "koukai/kenkyu jyugyo" and "jyugyo kentoukai". The process of transforming a planned curriculum, such as that found in National Course of Study or textbooks, into a curriculum that can be implemented in the classroom is referred to as "preparation", the first stage of the Lesson Study process. This process begins with finding and selecting materials relevant to the purpose of the class, and is then followed by refining the class design based on the actual needs of the students and tying all of this information together into a lesson plan. The significance of Lesson Study is that all of these processes are performed in collaboration with other teachers. A class (Photo 1) is then taught based on the teaching plan devised. The class is observed by many teachers, who are sometimes joined by university instructors and supervisors from the board of education, and a review session is held for all observers after the class. This process is shown in Figure 1. Steps (1) to (4) comprise the first stage, and the results of the evaluation in step (4) are utilized in the second stage, steps (5) to (7), to refine the class. The

Photo 1: A study lesson class with observers

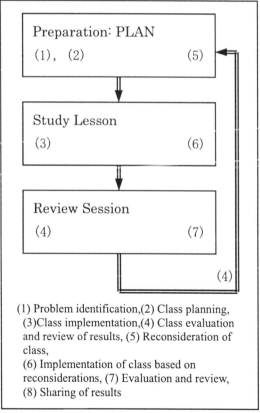

(1) Problem identification,(2) Class planning,
(3)Class implementation,(4) Class evaluation
and review of results, (5) Reconsideration of
class,
(6) Implementation of class based on
reconsiderations, (7) Evaluation and review,
(8) Sharing of results

Figure 1: Flowchart of Pedagogical Training (Stigler&Hiebert,1999)

thoughts of individual teachers, improvement of the level of teaching techniques, and the breadth of the network among teachers all come into play in this process.

2. Class and Discussion Topics: A Case Study

The following details a case study of a research classroom (Figure 2) to be evaluated. Given the introduction of new perspectives on scholastic ability, the current National Course of Study emphasizes the ability to think proactively and autonomously. For example, the question of how to incorporate the cultivation of proactive thinking in subject-based learning is an important practical study theme. Preparation is performed on this theme and a lesson plan is prepared. In some cases teachers are asked to develop an index for measuring the student's level of achievement by using specific numerical scores.

In the review session following the class, the instructor gives brief introductions and explains the purpose of the class. Based on the teaching plan distributed ahead of time, concepts on teaching materials and characteristics/status of the students are described in accordance with each stage of the class. Also, the rationales for each problem and activity conducted in the class are explained. Then each participant draws upon their own teaching experiences to express opinions and ask questions about the problems given in the class and teacher's instructive role, as well as about the students' utterances and learning activities. The purpose of this review session is to explore ways to improve the class by analyzing any disparities between the original goals established for the classroom or the plans developed to achieve those goals, and what is actually happening in the classroom (Figure 3). Of significant interest is the ability of this process to facilitate the discovery of new problems or issues that had not initially been noticed during the class.

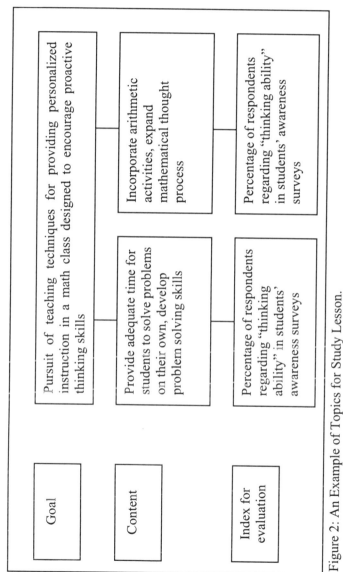

Goal	Pursuit of teaching techniques for providing personalized instruction in a math class designed to encourage proactive thinking skills
Content	Provide adequate time for students to solve problems on their own, develop problem solving skills	Incorporate arithmetic activities, expand mathematical thought process
Index for evaluation	Percentage of respondents regarding "thinking ability" in students' awareness surveys	Percentage of respondents regarding "thinking ability" in students' awareness surveys

Figure 2: An Example of Topics for Study Lesson.
(Edited excerpt from www.pref.hiroshima.jp/kyouiku/hotline/)

Types of Lesson Study Formats

Conducting Lesson Study involves a large number of teachers, but it can be done on many different scales and in varying formats (Table 1). The most common format is in-school training conducted at the school level. Lesson Study is conducted by developing an annual pedagogical theme and forming teams for each subject and grade. The classes themselves are actually taught by their regular teachers, but the process of working together empowers individual teachers in their classes and also fosters good relationships between colleagues who teach at the same school.

Other formats include Lesson Study conducted by groups of teachers on a voluntary basis and Lesson Study sessions hosted by teachers' unions and academic societies. Given the multitude of formats in which these efforts are being conducted, it appears that Lesson Study has taken root in Japan's educational culture in the true sense and that it is exerting a significant impact on the quality of education.

Finally, we will look at the systems that exist to provide public support for Lesson Study efforts. Teachers are legally and socially expected to continue upgrading their skills even after they have attained a teaching position. Public training workshops (new teacher training, annual training, etc.) are designed to give teachers opportunities to work on upgrading their skills based on their experience. Some workshops are mandatory while others are voluntary. Even in these training workshops, Lesson Study is employed as a strategy for cultivating teaching skills. When thinking about the global issue of improving the quality of education, Lesson Study, which has a dialectical relationship with the theories and practices employed on the front lines of education, must be versatile enough to be applicable beyond the Japanese context.

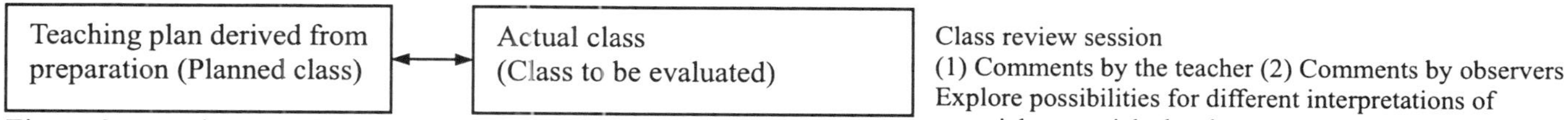

Figure 3: A review Session

Table 1: Lesson Study Formats in Japan (Ikeda et al., 2002, p. 28)

	Scale of participation	Main sponsor(s)
1	Individual schools	Public school principals and teachers (In-school training)
2	Study groups at the prefectural, municipal, and ward levels	Public school teachers
3	Study groups at the prefectural, municipal, and ward levels	Board of education, education center
4	Nationwide	Principals and teachers at schools affiliated with universities
5	Prefectural level, nationwide	Private institutions (academic societies, corporations, etc.)

Example: Hiroshima Prefecture
Implemented by: Board of Education, Education Center
· Compulsory training (1st and 2nd years, 6th year, 11th year)
· Voluntary training
Purpose: To improve teaching skills and problem-solving skills

Figure 4: Public training system (Edited excerpt from http://pfrq3.hiroshima-c.ed.jp/)

A Brief History of Mathematics Lesson Study in Japan

Section 2.1: "Where did Lesson Study Begin, and How Far Has It Come?"

Masami Isoda

Lesson Study began in the late 19th century with class visits designed to allow the whole classroom instruction.

1. From Individualized Instruction to whole classroom instruction: Studying Teaching Methods

Under the seclusion policies and class system that characterized the Edo period for about 260 years prior to the installation of the new Meiji government in 1868, literacy (and numeracy) education was available to commoners through *terakoya*, or temple schools, that had opened up autonomously around the country. Commerce thrived and the class system gradually collapsed during this period of seclusion, and by the late Edo period, individual knowledge and skills were highly regarded in the recruitment of workers. Due to the widespread emergence of temple schools, to which parents could voluntarily send their children, the literacy rate at the end of the Edo period was 43% among males and 10% among females, even then making Japan one of the most educated countries in the world. Individualized instruction was the common teaching method employed.

In 1872, the Meiji government issued the Education Code and at the same time established a teachers' school (normal school) in Tokyo (forebear to University of Tsukuba). With the goal of disseminating Western scholarship, the government invited foreign

Pestalozzi: Experience-oriented teaching methods

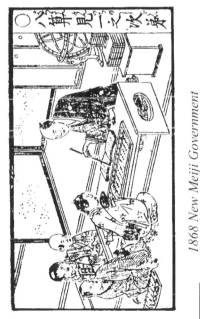

1868 New Meiji Government
Imperial government, country opening, Westernization policies

1872 Education Code issued, teachers' schools established
How is teaching done in a classroom?
How does group instruction work?

Figure 1: Shift from the curriculum and teaching methods of the *terakoya* (temple schools) to those of new types of schools.

teachers to teach Western subjects. The foreign teachers introduced the concept of whole classroom instruction, a style then still rare even in the West, into the teachers' school (Figure 1). The Japanese teachers and students, who were familiar only with the individualized instruction model in which subjects were taught individually based on the academic abilities of the student, learned not only the contents of the subject, but also methods of teaching by observing their teachers' behavior.

Textbooks created by foreign teachers at the teachers' school contained drawings of students raising their hands to answer questions posed by the teacher, as shown in Figure 2. It contained the question "How many students are raising their hands?" This foreign teacher wrote a textbook that teaches instruction methods as well as mathematics at the same time. The group instruction model implemented at the teachers' school in Tokyo spread to other teachers' schools around the country. Due to financial difficulties, the new government closed down all the teachers' schools except the one in Tokyo around 1880.

Nonetheless, in the decade while the schools were open, the practice of group instruction was disseminated around the country by graduates of the teachers' schools and by scroll pictures (Figure 1, right) and textbooks (Figure 2, right).

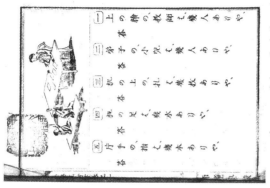

"How many people are raising their hands?"

Illustration from an elementary mathematics textbook in 1873.

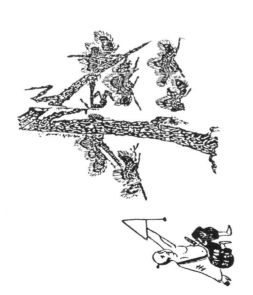

"How tall is the tree?"

Illustration from *Jinkoki*, a mathematics textbook from the Edo period.

Figure 2: From textbooks (left) that allowed students to study numeracy at their own discretion, depending on their needs, to textbooks (right) designed to allow students/teachers to simultaneously study leaning/teaching methods.

2. Dissemination of the Lesson Study Practices through the Elementary School Attached to the Tokyo Teachers' School.

In the 1880s, study on group instruction and its dissemination reached new heights as overseas study missions began returning to Japan. Mission delegates, who had been teachers at the teachers' school before their departures, were invited to become teachers at the elementary school attached to the teachers' school after their return, and a book on the Pestalozzi's teaching method was published. Even back then, this book contained comments on teaching materials, as well as instructions for conducting class observation and holding critique sessions. At the instruction of the Ministry of Education, these teaching methods were implemented throughout Japan as a model. Open classes, the origin of study lessons, were held to encourage the proposal of new teaching methods and teaching curricula, producing the first interactive Lesson Study groups initiated by the government. Figure 3 shows one of the national teachers' training conferences, which have been held since the Meiji period.

3. Development and Dissemination of Teaching Methods Learned through Lesson Study

As the country grew wealthier, it became possible for anyone to attend and graduate from elementary school. In the 1920s, new teaching methods based on the educational philosophy of scholars like John Dewey launched an era in which non-government-attached-school teachers began proposing their own teaching methods. At this time, a new teaching method was proposed for enhancing peer learning (see Figure 4). It allowed students to come up with their own study questions, discuss with one another whose question they wanted to research, and then go about researching the selected question. This set the stage for the emergence of teaching methods that focus on problem-solving, which today are globally recognized

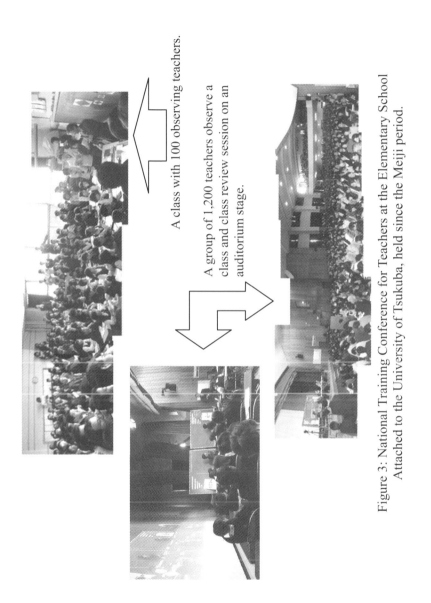

A class with 100 observing teachers.

A group of 1,200 teachers observe a class and class review session on an auditorium stage.

Figure 3: National Training Conference for Teachers at the Elementary School Attached to the University of Tsukuba, held since the Meiji period.

as models of constructivist teaching. Teachers' unions were launched after World War II, and Lesson Study by involved teachers led to heated debates. These classes also came to be used for launching futile ideological opposition. Teaching methods focused on problem-solving, which recognized the limitations of what already known and tried to produce new knowledge, were able to achieve success in spite of having to overcome the conflicts and other challenges. This was possible because visiting teachers were exposed to classes conducted for observation, and were impressed by seeing the students learning by themselves through problem solving.

Now, problem-solving approach is well known as a major way of teaching mathematics in Japan.

4. How Japanese Lesson Studies and Approaches are developed and known: A case of Open Approach

After 1943, Japanese National Secondary Mathematics Textbooks integrated different mathematical subjects into a single subject. Shigeru Shimada was an author. The textbooks were written with a focus on the processes of Mathematization and Open-ended problem solving. In the later 1960s, Shimada developed the research project of evaluation with Open-ended problems. In 1970s, the project had expanded to the Lesson Study projects for developing new teaching approaches; currently we distinguish them as 'Processes are open (various solving ways)', 'Ends are open (various answers against an open-ended problem)', and 'Problems are open (changing and developing problems from a problem)'. Later, Nobuhiko Nohda integrated them as the teaching method of 'Open Approach'. In the 1980s, Jerry Becker, Tatsuro Miwa and others began the collaborative study on problem solving between USA and Japan. On the contribution by Jerry Becker with co-researchers, as well as other simultaneous research movements, these are well known in the USA with classroom Lesson Study activities (See Chapter 5, Case 4 by Yoshihiko Hashimoto in this volume and Jerry Becker & Shigeru Shimada, 1997).

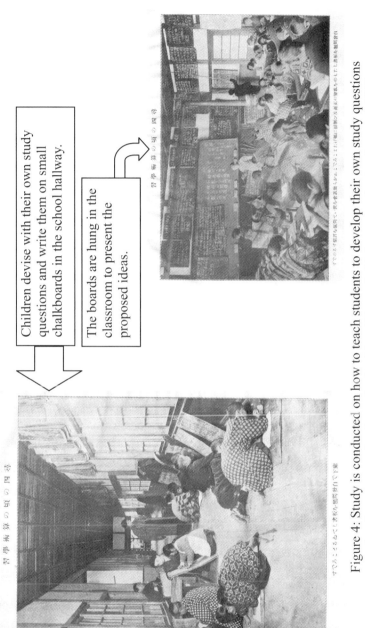

Children devise with their own study questions and write them on small chalkboards in the school hallway.

The boards are hung in the classroom to present the proposed ideas.

Figure 4: Study is conducted on how to teach students to develop their own study questions at the elementary school attached to Nara Women's higher normal school around 1920.

Official In-Service Teacher Training System

Section 3.1: How is In-Service Teacher Training Conducted in Japan?

Kazuyoshi Okubo

1. The In-Service Teacher Training System

Japanese education policies are aimed at developing people of well-rounded character with the ability to learn and think on their own, make decisions, act independently, solve problems, collaborating well with others, and to have compassion and sensitivity toward others. Achieving these goals hinges on the capabilities of teachers. The process of educating, hiring and training teachers presents an opportunity to improve teachers' capabilities (Figure 1). Training can be divided into three types, based on their relationship to the teachers' official duties. The first is training conducted by the government as part of the teachers' duties. The second is training that is independently conducted outside school during working hours and recognized under the Law for Special Regulations Concerning Educational Public Service Personnel. The third is voluntary training held outside working hours. Boards of education and universities with teacher certification programs support independent study groups in their local areas and offer venues for work-time training and voluntary training. Work-time and voluntary training programs tend to be organized by the teachers themselves and have the characteristics of "research."

2. Training Organized by the Government

The government holds planned, organized, systematic training programs. To achieve its educational goals, the national government provides financial support for training programs, and in each prefecture conducts training for teacher leaders and training to address pressing issues relating to school education (Figure 2). The content of the training is determined by officers (as well as councils,

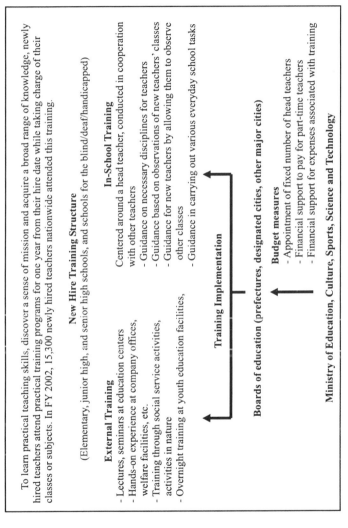

Figure 1: New Hire Training

etc.) assigned to research subjects. Supervisors of the prefectural education centers and other educational institutions plan and implement government training programs. Training for new teachers aims to impart practical teaching skills and a sense of mission, as well as to enable teachers to acquire a broad range of knowledge. Training sessions are attended for one year from the teachers' date of hire (see Figure 3). There is also 10th year training. It aims to improve teaching skills in various subjects and is adapted to the capabilities and needs of individual teachers. In addition, the education centers also provide special training classes to provide instruction in various subjects for teachers who want additional training. Special training courses are also held to teach new contents of a subject when education reforms are implemented, and to teach new evaluation methods when evaluation standards are revised.

3. In-School Training (Research)

In-school training forms the central role in the development of teaching skills. Administrators and teachers alike refer to training as "research." In particular, administrators promote in-school research to enable groups of teachers to systematically work with one another to improve their teaching skills by establishing a training committee and assigning training leaders for each grade. The training committee decides on the training topics the school will address for the following year and establishes a training plan for enhancing teachers' knowledge of those topics. For example, the committee determines who will conduct the classes that will be observed for evaluation each month. The assigned teacher then works either with a group of teachers of the same grade or with relevant teachers in other grades to plan and implement the class. Other interested teachers observe this class, and the person who is assigned to teach the observed class next uses the feedback obtained to plan their lesson. To ensure that the research goes smoothly, the training committee may organize training sessions led by university professors. The administrators provide external training opportunities to each teacher, based on the plan developed by the training committee. Results of in-school training are verified by what students are seen doing during classroom visits. Public

Financial Measures for Retaining Teachers

To ensure the retention of the most outstanding teachers, special measures even beyond those in place for general government workers, have been taken with regard to public school teachers' salaries. Half of the salaries of public elementary and junior high school teachers is paid for by the prefectural governments, while the other half is paid for by the national government (pursuant to the system where the national treasury pays a share of compulsory education expenditures).

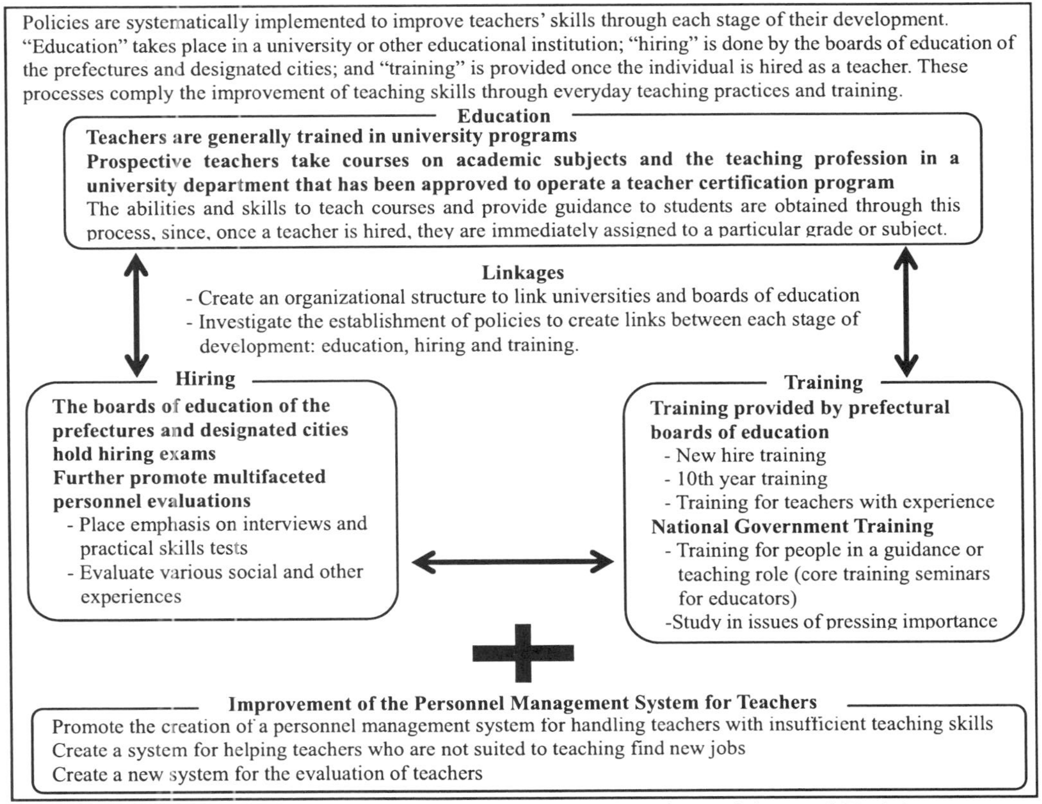

Policies are systematically implemented to improve teachers' skills through each stage of their development. "Education" takes place in a university or other educational institution; "hiring" is done by the boards of education of the prefectures and designated cities; and "training" is provided once the individual is hired as a teacher. These processes comply the improvement of teaching skills through everyday teaching practices and training.

Education

Teachers are generally trained in university programs

Prospective teachers take courses on **academic subjects and the teaching profession in a university department that has been approved to operate a teacher certification program**

The abilities and skills to teach courses and provide guidance to students are obtained through this process, since, once a teacher is hired, they are immediately assigned to a particular grade or subject.

Linkages

- Create an organizational structure to link universities and boards of education
- Investigate the establishment of policies to create links between each stage of development: education, hiring and training.

Training

Training provided by prefectural boards of education
- New hire training
- 10th year training
- Training for teachers with experience

National Government Training
- Training for people in a guidance or teaching role (core training seminars for educators)
- Study in issues of pressing importance

Hiring

The boards of education of the prefectures and designated cities hold hiring exams

Further promote multifaceted personnel evaluations
- Place emphasis on interviews and practical skills tests
- Evaluate various social and other experiences

Improvement of the Personnel Management System for Teachers

Promote the creation of a personnel management system for handling teachers with insufficient teaching skills
Create a system for helping teachers who are not suited to teaching find new jobs
Create a new system for the evaluation of teachers

Figure 2: Systematic Structure Throughout Education, Hiring and Training

research meetings are occasionally held to provide teachers with a forum for presenting the achievements of their training.

4. Voluntary Training and Work-time Training

Voluntary and work-time training programs that do not receive financial support are carried out through teachers' voluntary efforts. One way they can do this is to participate in public research meetings held by other schools. Observing well-conducted classes and learning stimulates teachers and helps them develop better classes. If government training is a top-down approach to training, these kinds of training represent a teacher-instigated bottom-up approach. Achievements of training are shared with the public through national conferences of academic societies (such as the Japan Society of Mathematical Education) and at meetings for presenting research that are held by mathematical education associations at the prefectural and municipal levels.

5. A Wider Perspective

These three categories – Training organized by the government, In-school training (Research), and Voluntary training and Wok-time training – reflect either a management or a perspective based on financial responsibility for Lesson Study activities. While this is a useful perspective, Lesson Study does need to be seen from the perspective of teachers themselves. Teachers see Lesson Study as an integral part of their professional life regardless of who is organizing it, and as something to be enjoyed.

Teachers in Japan expect to participate in Lesson Study activities at all stages of their careers and make different contributions as they become more experienced or take on higher responsibilities in schools. During their careers, they are expected to deepen through their own practice and participation in Lesson Study their knowledge of students' development and how to cultivate rich learning. In addition, many other agencies participate in and support Lesson Study activities – universities, publishers and so on. Their roles have been discussed chapter 4 in this book and in the Preface.

– Training –

1 Enhancement of the Systematic Training Structure

To fulfill their responsibilities, educators are required to undergo continuous training. The boards of education of prefectures, designated cities and other major cities are required to offer planned training sessions. A systematic training structure has been developed for different types of training starting with new hire training.

The national government provides support for the training activities implemented by the prefectural governments and also holds training sessions for educators who are in leadership roles and seminars on pressing topics relevant to school education at the National Center for Teachers' Development.

Teacher Training Structure

	1st yr.	5th yr.	10th yr.	15th yr.	20th yr.	25th yr.	30th yr.	35th yr.

National level training / Leadership training:

Core Training Courses for Educators
For leader teachers ⟶ for principals and vice-principals

Seminars on giving guidance to students on courses after graduation, seminars for teachers of new industrial technologies

Overseas delegations of teachers
Short-term overseas delegations

Overseas training for young teachers
Japan-U.S. exchange programs
Delegations of young teachers to the U.S.

Seminars on pressing topics	Training for teachers promoting the use of information technologies in education, training sessions on education to prevent AIDS and drug abuse

Training implemented by boards of education of prefectures, designated cities, and core cities:

New hire training   10th year training
Training for teachers with experience
5th year training   20th year training

Training for student guidance supervisors
Training for new head teacher of instruction
Vice-principal training
Principal training

Long-term external training at a private company

Specialized training in subject teaching, student guidance (held by education centers)
In addition to these, voluntary training sessions are also held at municipal board of education facilities and in schools.

Figure 3: Improve Teaching Skills, Expand Horizons

Note: Figures 1-3 are excerpts from the "Pursuit of Teachers with Great Characters" article on the website of the Ministry of Education, Culture, Sports, Science and Technology.
http://www.mext.go.jp

Mathematics Curriculum and Way of Implementation

Section 4.1: How Has Mathematics Education Changed in Japan?

Eizo Nagasaki

Mathematics education in Japan since the Meiji period, when the shift was made to a modern education system, can be divided into five phases from a curriculum development perspective:

Phase I: Assimilation of mathematics education from western Europe (1860s to 1930s)

Phase II: Formation of Japan's own mathematics education (1930s to 1940s)

Phase III: Establishment of the foundations of Japanese mathematics education (1950s to 1960s)

Phase IV: Modernization of mathematics education based on international trends (1960s to 1970s)

Phase V: Development of mathematics education suitable for students (late 1970s forward)

Phase I began with the use of mathematics curricula from other countries and translated textbooks. Ideas for mathematics reforms proposed by such foreign mathematicians as John Perry and Felix Klein were later compiled into the "Mathematics Education Improvement Campaign" and these led to improvements in Japanese mathematics education.

In Phase II, there was a shift in mathematics education from the perspective of "teaching mathematics in a way that makes it easy to understand" to the perspective of "teaching children to create

Table 1: Changes in Mathematics Education in Japan and Around the World (continue to page 25)

Trends in Japanese mathematics education	World trends as they pertain to Japan
I. Assimilation of mathematics education from western Europe **1860s-1930s** 1868 Meiji Restoration. 1872 Regulations for the National Courses of Study for Elementary and Junior High Schools (first national curricular standards). 1905 A textbook for elementary schools, *Ordinary Elementary School Arithmetic* (black cover), was first used (the first indigenous Japanese elementary school mathematics textbook). 1919 The mathematical Association of Japan for Secondary Education (precursor to the Japan Society of Mathematical Education). 1925 Metric system begins to be incorporated into mathematics textbooks. 1931 National courses of study for junior high school mathematics are revised.	1901 Perry lectures "Teaching of Mathematics" in England. 1902 Secondary education system reforms are implemented in France. 1902 R. L. Moore lectures on the foundations of mathematics in USA. 1904 Felix Klein lectures "Mathematics Education in High School" in Germany. 1912 Japan gives a presentation at a meeting of the International Commission on the teaching of Mathematics at the 5th International Congress of Mathematicians (ICM).
II. Formation of Japan's own mathematics education **1930s-1940s** 1935 Elementary schools begin to use *Ordinary Elementary School Arithmetic* (green cover), a textbook for elementary schools. 1940 Committee for Reorganizing Syllabi of Mathematics is established. 1942 National Courses of Study for junior high schools and girls' high schools are revised. 1942 Schools begin *First and Second Categories in Mathematics*, a junior high school textbook.	1936 Japan gives a presentation at a meeting of the International Commission on the teaching of Mathematics at the 10th International Congress of Mathematicians (ICM). 1938 American Progressive Education Association publishes *Mathematics in General Education.*
III. Establishment of foundations of Japanese mathematics education **1950s-1960s** 1951 The National Course of Study (Tentative) is published for elementary, junior high, and senior high schools (learning by the method) 1953 JSME holds the Joint Conference on Curricula of Elementary, Junior High, and High Schools. 1955-60 The National Course of Study for elementary, junior high, and senior	1951 University of Illinois Committee on School Mathematics (UICSM) is launched. 1956 Japan participates in the International Conference on Public Education, Mathematics Education in Higher Education, under

mathematics." [1] However, this was never able to be fully implemented because of the war.

Phase III began after the war with "Learning by Unit", which emphasized the social utility of mathematics[2], but a movement by the Japan Society of Mathematical Education (JSME) later led to a shift toward systematic learning, which emphasized the structures of mathematics. It was this shift that firmly established the foundations of Japanese mathematics education.

In Phase IV, Japan took lessons from the modernization of mathematics education in other countries and, spurred by a movement of the JSME, introduced modern concepts and approaches into Japanese mathematics education with the terminologies for developing mathematical thinking.

In Phase V, international cooperation in the form of international studies and conferences has flourished, and efforts are being taken to develop mathematics education designed to meet the needs of primary and secondary school students.

---

[1] Editors' note: Mathematical activities were enhanced from this age on the national curriculum.

[2] Editors' note: It was the second influence of the progressivism from US.

Table 1 (continued)

	UNESCO.
high schools is issued (systematic learning).	
IV. Modernization of mathematics education based on international trends 1960s-1970s	1958 School Mathematics Study Group (SMSG) is launched in the USA.
1961 JSME publishes the Research Journal of Mathematical Education	1959 Organization for European Economic Cooperation (OEEC) holds a seminar on "New Approaches to Mathematics."
1963 JSME launches the Mathematics Curriculum Research Committee.	1964 International Association for the Evaluation of Educational Achievement (IEA) holds its First International Mathematics Study.
1966 JSME publishes *The Modernization of Mathematics Education.*	1969 The 1st International Congress on Mathematical Education (ICME) is held (Lyon, France).
1968-70 The National Course of Study for elementary, junior high, and senior high schools is issued (modernization).	1974 ICMI-JSME Regional Conference on Mathematical Education is held.
V. Development of mathematics education suitable for students Late 1970s and forward	1980 National Council of Teachers of Mathematics (NCTM), USA, announces its "Agenda for Action" (emphasizing problem solving).
1977-78 The National Course of Study for elementary, junior high, and senior high schools is issued (fundamentals/basics).	1980 IEA holds its Second International Mathematics Study.
1989 The National Course of Study for elementary, junior high, and senior high schools is issued (internationalization/computerization/individualization).	1983 ICMI-JSME Regional Conference on Mathematical Education is held.
1998-99 National Course of Study for elementary, junior high, and senior high schools is issued (zest for life).	1990- NCTM, USA, announces "Curriculum and Evaluation Standards for School Mathematics."
2002 The National Institute for Educational Policy Research conducts a Study of Curriculum Implementation Status among elementary, junior high, and senior high school students.	2000- The 9th International Congress on Mathematical Education (ICME) is held (Tokyo).
	2000- Organization for Economic Co-operation and Development (OECD) implements Programme for International Student Assessment (PISA).

Section 4.2: How Have the Goals of the Mathematics Curriculum Changed?

Eizo Nagasaki

The national curricular standards for elementary, junior high, and senior high schools in Japan in the postwar period are stipulated in the National Course of Study (*gakusyu shidou youryo*). The changes in the goals of mathematics education can be understood by looking at the mathematics goals stated in the National Course of Study in each decade. The goals are described in four evaluation perspectives in the Permanent Cumulative Record (*gakusyu shidou youroku*) that provides national standards for developing student evaluations' criteria. The table below shows features of the curriculum, the mathematics goals of the National Course of Study for elementary schools, and the criteria of mathematics evaluations in the Elementary School Permanent Cumulative Record.

The goals of the mathematics curriculum in Japan have included the "cultivation of mathematical thinking" since the 1960s.[1] Other mathematics goals in each decade are as follows:

1950s: Education in mathematics problem solving skills to solve social problems

1960s: Understanding of mathematics concepts

1970s: Cultivation of the ability to think from integrating and developing point of view.[2]

---

[1] Editors' note: The cultivation of mathematical thinking has being enhanced for clarifying the quality of mathematical activities which were enhanced in 1950s.

[2] Editors' note: It used to be focusing on the restructuring nature of mathematical development with the invariant terminologies of mathematical

Table 1: Changing Goals of the Mathematics Curriculum (continue to page 29)

Decade	Major features of the mathematics curriculum	Mathematics goals of the National Course of Study for elementary schools	Criteria of mathematics evaluations in the Elementary School Cumulative Guideline Reports
1950s	**Learning by the unit method** Curriculum consists of mathematics concepts and methods for understanding and finding mathematical solutions to social problems and problems concerning everyday issues.	(1) It is important to improve skills that enable students to solve the problems that arise in everyday life on their own, as necessary. (2) It is important for children to have the desire to make their lives better through quantitative processing. (Remainder omitted.)	Interest/attitude toward numbers and quantity Quantitative insight Logical thinking Calculation and measurement skills
1960s	**Systematic learning** Curriculum consists of mathematics concepts and mathematical thinking.	1. Enable children to understand the basic concepts and principles of numbers and figures, and foster more advanced mathematical approaches and ways of processing information. 2. Teach basic knowledge pertaining to numbers and figures as well as proficiency in basic skills, and enable children to use those skills accurately and efficiently for a given purpose. (Remainder omitted.)	Interest in numbers and quantity Mathematical thinking Understanding of terms and symbols Calculation skills
1970s	**Modernization** Curriculum incorporates modern mathematics approaches, concepts, and content.	Cultivate skills and attitudes that will allow students to have a mathematical understanding of everyday events, think coherently, and to observe and process information in an integrated and developmental manner. (Remainder omitted.)	Knowledge/understanding Skills Mathematical thinking

1980s: Acquisition of basic mathematics knowledge and skills
1990s: Appreciation of the meaning of mathematics
2000s: Getting enjoyment from mathematics activities

Based on this framework, one could argue that the goals of mathematics education in Japan emphasized the social need for mathematics in the 1950s, the mathematical need for mathematics in the 1960s and 1970s, and the needs of children since the 1980s.

The more recent goals of mathematics education in Japan, analytically speaking, have consisted of four main components: interest / eagerness / attitude, mathematical thinking, expression / processing, and knowledge / understanding.

Editors' Note: Further readings

Editorial Department of Japan Society of Mathematical Education (JSME) published the following books in English on mathematics education in 2000: 1) Mathematics Education in Japan during the Fifty-five Years since the War: Looking towards the 21st Century. 2) Exploring Elementary School Mathematics Education of Japan in the 21st Century; Based on Practical Studies in the 1990s. 3) Exploring Secondary School Mathematics Education of Japan in the 21st Century; Based on Practical Studies in the 1990s. 4) Mathematics Program in Japan; Elementary, Lower Secondary & Upper Secondary Schools.

Further information:
http://www.sme.or.jp/e_index.html

---

thinking. The ideas originated in Japan by World War II, and are not the same but not so far from *Mathematization* by H. Freudenthal (1968/1973).

Table 1 (continued)

1980s	**Fundamentals/basics** Curriculum stripped of modern mathematics concepts and content that were introduced during the period of modernization.	Impart fundamental knowledge and skills pertaining to numbers and figures, cultivate skills and attitudes that will allow students to have a mathematical understanding of everyday events, think and process information coherently.	Knowledge/ understanding Skills Mathematical thinking Interest/attitude toward numbers and figures
1990s	**Internationalization/computerization /individualization** Curriculum actively incorporates computer and elective learning.	Impart fundamental knowledge and skills pertaining to numbers and figures, cultivate skills that will allow students to have an outlook on everyday events and to think coherently, understand the benefits of mathematical processing, and cultivate a positive attitude toward the voluntary use of mathematics in everyday life.	Interest/eagerness/attitude toward mathematics Mathematical thinking Expression/processing of numbers and figures Knowledge/understanding of numbers and figures
2000s	**Zest for life** Curriculum demonstrates greater selectivity of mathematics content and emphasizes mathematical activities of students while expanding the elective learning structure.	Impart fundamental knowledge and skills pertaining to numbers and figures through mathematical activities, cultivate skills that will allow students to have an outlook on everyday events and to think coherently, impart the fun and benefits of mathematical processing, and cultivate a positive attitude toward the voluntary use of mathematics in everyday life.	Interest/eagerness/attitude toward mathematics Mathematical thinking Expression/processing of numbers and figures Knowledge/understanding of numbers and figures

Section 4.3: How are Curriculum Standards Improved and Implemented?

Yutaka Ohara

In Japan, schools develop their own education curricula based on standards established by the Ministry of Education, Culture, Sports, Science and Technology. Lesson Study is conducted either to make suggestions for the implementation of the curriculum, or to revise the curriculum.

1. Improving Curriculum Standards

Curriculum standards are comprised of the National Course of Study, which establishes the how schools are to develop their curricula, and the goals and content to be covered in each subject and grade level, and the School Education Law Enforcement Regulation, which establishes the number of required class hours. Curriculum standards are revised in approximately ten year cycles, and a transition period of three years has been allotted from the time any improvements are announced to the time of their full implementation. Revisions to the National Course of Study and School Education Law Enforcement Regulation are made in accordance with official procedures. First a report is compiled by a committee comprised of designated experts (Central Council for Education) that confirm the basic principles of the improvements that are to be made. This is followed by a report by a committee that confirms the revision guidelines and the class hours. Finally, a committee is convened to confirms the subject matter and teaching systems for each subject. Up until the most recent revisions, follow-up committees were going to be set up based on the results of the deliberations of these upper level committees, but in 2004 a

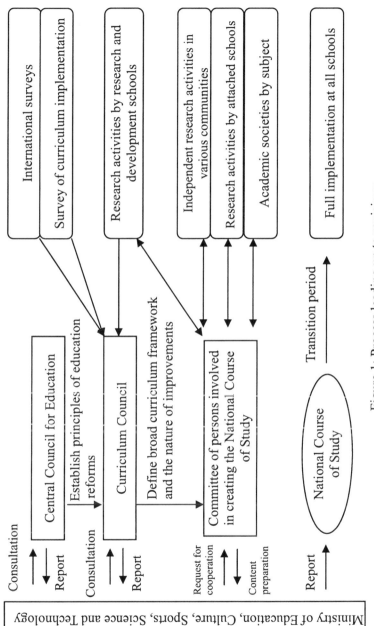

Figure 1: Process leading up to revisions

new format was adopted. Now a committee has been permanently established to fulfill the roles by these committees which will proceed with revision preparations concurrently.

The process of summarizing the objectives and subject matter to be covered in each subject, revising curriculum standards, and disseminating new information is overseen by MEXT's school inspector, school subject investigator, and other cooperative people. MEXT and the National Institute for Educational Policy Research (NIER) Curriculum Research Center perform various tasks related to implementing the curriculum in each subject and evaluating the status of implementation. For example, they conduct curriculum implementation status surveys during the transition period and after improvements have been made, issue recommendations for improving teaching methods to meet curriculum standards, and prepare revisions to the curriculum standards themselves. The school subject investigator promotes progressive Lesson Study on the revision and implementation of standards at designated research schools, and promotes improvements to standards based on proposals made by academic societies and boards of education.

2. Curriculum System at Each School

No matter how good a curriculum is developed, it has little meaning unless it is actually implemented. There are three types of educational curricula: the intended curriculum, the implemented curriculum, and the achieved curriculum. Policies need to be enacted to ensure that the intended curriculum is implemented and that the aim of that curriculum, student growth, is achieved. As a basis for implementing these policies, Japan has a system of educational laws governing these three types of curricula.

Reference materials to be used by individual schools when developing their own evaluation standards and methods.

(1) Explanation of National Curriculum Guidelines (MEXT)

A guide provided to explain and disseminate the key points and nature of revisions to the curriculum guidelines.

(2) Materials Pertaining to Individual-Oriented Instruction (MEXT)

Reference materials for teachers on each school's efforts to improve teaching strategies for promoting developmental learning and supplemental learning.

(3) Report on the Survey of Curriculum Implementation Status (NIER Curriculum Research Center)

A survey report that helps improve teaching by facilitating an understanding of how lessons based on the objectives and subject matter of each subject are implemented in the context of school curricula developed in accordance with the National Curriculum Guidelines.

(4) Reference Materials for Creating Evaluation Standards and Improving Evaluation Methods (NIER Curriculum Research Center)

Figure 2: Instructional materials issued by the national government to ensure implementation of the curriculum standards (sample)

The intended curriculum is developed by the school principal in accordance with the curriculum standards. The implemented curriculum is controlled by teachers who are responsible for using textbooks developed in accordance with the curriculum standards in their classes. The achieved curriculum is monitored by the use of student report cards that record the students' results for the year.

School education in Japan utilizes a textbook investigation system in which items that meet the curriculum standards are adopted for use as textbooks. For compulsory levels of education, that is, elementary and junior high school, textbooks are distributed free of charge. Because of financial restrictions, the number of pages and colors that can be used in these textbooks are limited. Recently there has been a trend toward relaxing those restrictions. If we want to achieve a distinctive education and promote developmental learning, textbooks need to be more individualized. At most until 10% of the content of textbooks for compulsory grade levels[1] and 20% of textbooks for upper secondary schools should contain content that goes beyond the scope of the curriculum standards. Textbooks are typically revised three times while a single set of National Course of Study is in effect.

Under the current curriculum standards, the evaluation standards used for filling out the Permanent Cumulative Record have shifted from relative to absolute standards. It has to be kept at least five years but it is kept permanently. The NIER has proposed a set of evaluation standards that will ensure that the curriculum standards are being met in teaching.

---

[1] Editors' note: A current topic of Lesson Study in compulsory levels is *the developmental and supplemental learning* to integrate additional contents on the textbooks.

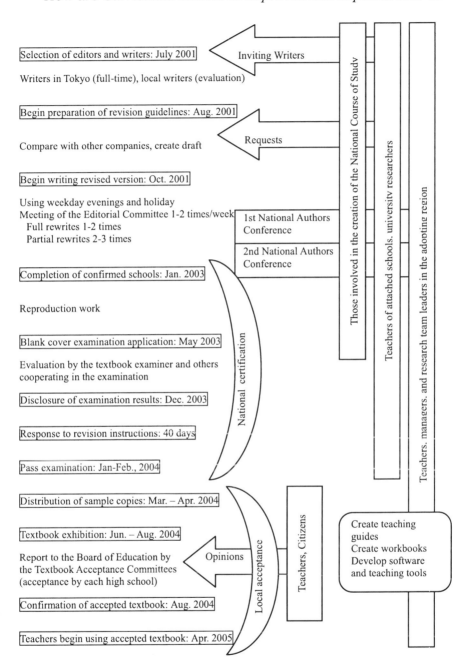

Figure 3: Sample process by which a textbook is revised and accepted

Section 4.4: How is Each School's Mathematics Curriculum Formulated and Implemented?

Shigeo Yoshikawa

The formulation and implementation of the curriculum is conducted internally within each school. Improvements are then made using feedback provided through the lesson study process.

1. Formulation of the School Curriculum

The school principal is responsible for developing the school's curriculum based on the National Course of Study, which embodies the curriculum standards, and the concerning laws and regulations. Each school creates an annual lesson plan and class schedule at the beginning of the school year based on the principal's instructions. The principal usually delegates the formulation activities to an internal department, like the educational affairs or research department of the school. The coordinating department oversees the annual teaching plan and class schedule created by the class teachers, teachers in charge of respective subjects, and the head teacher of each grade level. These plans are deliberated by the teachers' council, and the deliberation results are used by the principal to complete the formulation of the curriculum.

The general provisions of the National Course of Study confirm that the subject matter described in the plans are to meet certain minimum standards and require that each school formulate its own distinctive curriculum.To promote the development of a distinctive school and ensure that teachers are organized into independent and collaborative teaching groups, the principal prepares to put the right people in the right jobs starting in the previous school year and creates a system that facilitates his or her own ability to lead. As part of these preparations, the principal organizes internal departments like the educational affairs department or research department at the end of the previous school year. These departments cooperate on developing the school's unique

Evaluation Standards of the NIER Curriculum Research Center

Table 1: Internal Organization in a School

Department/coordinating body	1st Year Team	2nd Year Team	3rd Year Team	...
Educational Affairs Department: Planning and implementation of the curriculum	Teacher A (grade level officer, Japanese subject)	Teacher C (mathematics coordinator)	Teacher E (grade level officer)	
Research Department: Planning and implementation of teacher training	Teacher B (mathematics research leader)	Teacher D (grade level officer, Japanese subject)	Teacher F, mathematics research leader	
...				

Principal = vice principal = educational affairs head teacher (other than the class teacher)
= Teachers meeting (all teachers, called by the principal)
= Department meetings (held by the head teacher of each department
= Grade level meetings (held by head teacher of each grade level)
= Subject meetings (held by head teachers of each subject)

Announced to each school level (the following shows the case of an elementary school)
Chapter 1 General provisions (explanation of how to formulate a curriculum for that school level)
Chapter 2 Explanation of the subject matter to be taught in each subject
Section 3 Mathematics 3.1 Objectives
Impart fundamental knowledge and skills pertaining to numbers and figures through mathematical activities, cultivate skills that will allow students to have an outlook on everyday events and to think coherently, impart the fun and benefits of mathematical processing, and cultivate a positive attitude toward the continued use of mathematics in everyday life.
3.2 Goals and subject matter of each grade level
1st Year: 1. Objectives (explains the goals of each grade level)
1.1 Impart a strong sense of numbers through activities that make use of actual objects. (Remainder omitted.)
2. Subject matter (explains the subject matter for the grade level in each subject
A Numbers and Mathematics 1. Understand the meaning of and be able to use numbers through activities like counting the number of objects.
a. Compare the number of objects by performing correspondence operations. (Remainder omitted.)
Terms, symbols 3. Handling of the subject matter (each grade level)
3.3 Creation of teaching plans and the handling of subject matter across grade levels

Figure 1: Composition of the National Course of Study

curriculum (the educational goals of that particular school) and issues proposals to the principal and teachers. Through a meeting of the teachers' council, the principal asks each teacher to formulate a curriculum that achieves the distinctive goals of the school. If the principal follows the procedures for formulating the school's curriculum under the agreement of the teachers as a group, the curriculum for each class becomes curriculums that are determined by individual teachers. The implementation of the curriculum is the individual responsibility of each teacher as well as the collective responsibility of the teachers of the school as a group.

2. Preparation of an Annual Teaching Plan

The mathematics curriculum is created and implemented by the class teachers as their annual teaching plan. The National Course of Study, which stipulates the legally required subject matter, only provides basic guidelines regarding the objectives and subject matter to be covered in each grade level and the preparation of teaching plans. MEXT issues manuals and instructional materials to help teachers better understand the key points of those guidelines. Because annual teaching plans are created by the school, the National Course of Study outlines the subject matters to be taught in each grade level, but do not designate the order in which that information is to be taught. To respect the creation of a distinctive curriculum at each school, the subject matter for each grade level should be systematically planned in advance, within a range that is not too difficult for students.

Annual teaching plans stipulate the subject matter to be covered for the year and the teaching objectives. Teachers in charge of each class plan the school's objectives (the distinctive goals mentioned above) based on the National Course of Study, and then plan the subject matter to be taught, the order in which to teach it, and the objectives for the specific teaching term based on the goals of the teachers they oversee. This is an activity that relies heavily on the teacher's wishes and capabilities. MEXT and NIER publish materials and evaluation regulations for supporting the creation of

Table 2: Four Evaluation Perspectives in the form of statements that specify mathematics objectives

Interest, eagerness, attitude toward mathematics	Mathematical ways of thinking	Expression and processing of numbers and figures	Knowledge and understanding of numbers and figures
Has an interest in mathematical phenomena, appreciates the fun of the activities and the benefits of the mathematical process, seems willing to try to apply what he/she has learned to everyday phenomena.	Has acquired the fundamental mathematical ways of thinking through mathematical activities, establishes a plan of action with an outlook for what lies ahead.	Has acquired the skills involved in expressing and processing numbers and figures.	Has a strong sense of numbers and figures and understands their meaning and properties.

(1) A. Numbers and Calculations
Content of the National Course of Study
1 Understand the meaning of and be able to use numbers through activities like counting the number of objects.
a. Compare the number of objects by performing correspondence operations.
b. Correctly count and express a number of items and identify their order.
c. Create a sequence and express it on a number line by thinking about the size of numbers and their sequence.
d. Look at a number as the sum or difference of other numbers, assign relationships between numbers.
e. Understand the meaning of numbers up to 100 and how to express them.
2 Understand the meaning of addition and subtraction, and know how to use them.
a. Know when addition and subtraction are used, express those calculations as formulas, and read those formulas.
b. Think about the process of performing addition or subtraction calculations of two single-digit numbers, learn to perform those calculations properly.
3 Collect, count, and divide specific objects equally, organize and express those objects.

Figure 2: Evaluation standards for each subject matter (1st year)

well developed teaching plans and provide support for these activities.

3. Promoting Lesson Study and Determining Results

At the beginning of each school year, in addition to its annual teaching plan, each school establishes action plans, teacher training plans, a class schedule for each grade level, and the schedule for teachers' organizational activities. The teachers' organizational activities include weekly subject meetings for each academic subject, grade level meetings for each grade level, and educational affairs department meetings for each department. Participants in these meetings discuss the details of the curriculum implementation and make revisions to established plans. Training sessions for improving lessons based on objectives and examining student learning are held about 10 times a year in the form of research lessons. A teaching supervisor from the board of education is invited to observe these lessons, and often provides suggestions about whether teaching is being conducted appropriately based on the curriculum standards.

The lesson study results of each school are shared with others through magazines for teachers, organizational research reports issued by the board of education, and research meetings held by academic societies. MEXT and the NIER Curriculum Research Center study teaching practices and student achievement through visits to research and development schools, seminars by teaching consultants, and curriculum implementation surveys. Using this information to obtain clues for making improvements, they develop policies to help improve lessons. This is done by using MEXT publications and teaching materials that summarize the curriculum standards, and by working with cooperative individuals.

Table 3: Evaluation Standards for "A: Numbers and Calculations "

| Enjoys working with numbers and calculations and strives to gain various experiences with them. | Through mathematical activities like measuring the actual size and order of objects, and expressing their size in a chart or formula, student devises creative strategies and develops ways for expressing numbers and performing mathematics calculations. | Can easily perform simple addition and subtraction problems using whole numbers. | Has a strong sense of numbers, knows how to express the meaning of numbers, understands the meaning of addition and subtraction calculations performed on whole numbers (natural number and zero). |

Table 4: Sample evaluation standards for each subject matter (partial)

| Correctly counts and expresses the number of objects, such as concrete objects. Appreciates the benefits of using numbers to express the quantity or order of objects. Appreciates the benefit of being able to use numbers to know the size or sequence of numbers. | Through exercises involving numbers, knows how to read and express numbers and how to think about the size and sequence of numbers. | Can read and express numbers up to 100. Can correctly count and express the number of objects and identify their order. | Has a strong sense of the size and structure of numbers. Looks at a number as the sum or difference of other numbers, has a strong sense of the structure of numbers. Numbers up to 100. |

Section 4.5: Teaching and Assessment Based on Teaching Guides

Masao Tachibana

1. The Purpose of Assessment

Assessments are educational activities conducted for the purpose of helping teachers improve their lessons and for the better growth of students.

Assessments are not conducted for the mere sake of assessing, but to enable students and teachers to guide their learning situation in a positive direction and to reflect on their learning and teaching.

When students finish studying a particular topic, an assessment is carried out to see whether the goals of their activities were achieved. If the student's performance is not what was hoped, some teachers bemoan student's inattentiveness or feel disappointed in their students. This is not an appropriate response.

Since the primary goal of assessment is to ensure that teaching is well-suited to the students' learning capabilities, it is important that the teacher has a solid understanding of their students' learning status. Thus, activities to observe and measure students' capabilities through the use of tests and other means are often incorporated into assessments of students. Traditionally, assessments have tended to be thought of as a process of obtaining this kind of information and expressing it as a number or score. Given their objectives, however, assessments are significant in and of themselves as resources that can be used to help teachers and students examine and improve their own behavior.

2. Assessments that Improve the Teacher's Lessons

Conducting a lesson based on a goal makes it possible to evaluate the lesson, and through that evaluation, lessons will improve.

Relationship Among National Course of Study as Standards, Teaching Materials, and Journals

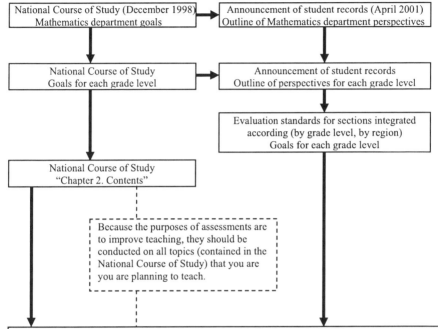

National Course of Study (December 1998)
Mathematics department goals

Announcement of student records (April 2001)
Outline of Mathematics department perspectives

National Course of Study
Goals for each grade level

Announcement of student records
Outline of perspectives for each grade level

Evaluation standards for sections integrated
according (by grade level, by region)
Goals for each grade level

National Course of Study
"Chapter 2. Contents"

Because the purposes of assessments are
to improve teaching, they should be
conducted on all topics (contained in the
National Course of Study) that you are
you are planning to teach.

National Institute for Educational Policy Research (NIER)
"Reference Materials for Creating Assessment Standards and Improving Assessment Methods"
"Specific Examples of Assessment Standards for Each Section Integrated According to Content"

The Reference Materials contain the "Specific Examples of Assessment Standards for sections integrated according to content" from four different perspectives that roughly correspond to the content in the National Course of Study. They indicate when a student is in a "generally satisfactory" situation and are written in the aim to achieve that situation and assess students of that level.

Teaching must occur for assessment to be performed. That is, "you cannot assess that which has not been taught."

Assessment standards consist of the goals of the teacher for raising the student to a generally satisfactory level and allow the student to understand what they are trying to accomplish in their learning. Thus, teachers need to use the assessment guidelines as a point of reference in thinking about how they can use their teaching materials to develop aimed abilities in the National Course of Study to their students.

Once the study of a particular topic has ended, it is important to identify, based on the goals of the lesson, how effectively the students learned, what kinds of issues they had trouble with and which students struggled with what content.

By identifying these things, teachers can examine whether the learning activities and a teacher's method of teaching and supporting students are effective for the achievement of the curricular goals. Depending on the situation, it is possible to revise the teaching plan, improve what needs to be improved, and enable students to achieve the goals of the curriculum.

Assessments should reflect the Mathematics goals stipulated in the National Course of Study and the goals and curricular content for the particular grade level. The Reference Guides for Creating Evaluation Standards and Improving Evaluation Methods issued in February 2002 by The National Institute for Educational Policy Research (NIER) were created based on the National Course of Study and the descriptions and outline of the viewpoint of those guidelines. Thus, they can be used as a reference for teaching and conducting assessments.

3. Assessment to Improve Student Learning

Assessment should act as a "message from the teacher" to help improve the student's academic ability.

When conducting an assessments of students' work, it is important that the teacher conveys evaluative information that is specific and easy to understand, such as, "This is how much you were able to do," "You started out a little slow in this area, but you worked hard and were able to improve," or "You'll do even better in the future if you pay particular attention to this." In using assessment, teachers must be careful to avoid pointing out a student's weaknesses or shortcomings, but encourage student learning by identifying areas where the student needs to expend more effort. Students who are not achieving their goals often do

Evaluation Standards for Teaching,
Standards for Student Assessment from Four Perspectives to Reflect and Foster Student Growth

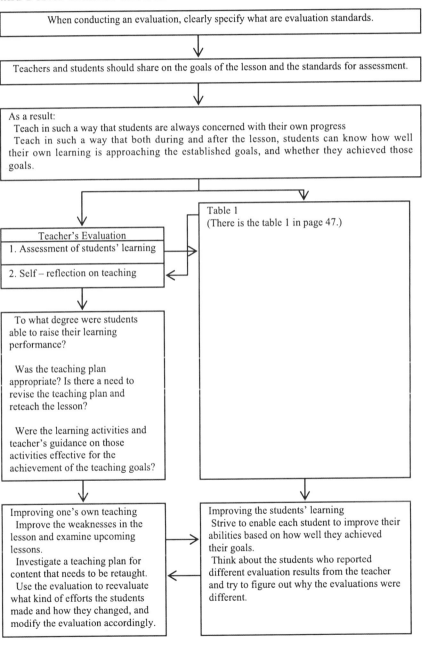

When conducting an evaluation, clearly specify what are evaluation standards.

Teachers and students should share on the goals of the lesson and the standards for assessment.

As a result:
Teach in such a way that students are always concerned with their own progress
Teach in such a way that both during and after the lesson, students can know how well their own learning is approaching the established goals, and whether they achieved those goals.

Table 1
(There is the table 1 in page 47.)

Teacher's Evaluation
1. Assessment of students' learning
2. Self – reflection on teaching

To what degree were students able to raise their learning performance?

Was the teaching plan appropriate? Is there a need to revise the teaching plan and reteach the lesson?

Were the learning activities and teacher's guidance on those activities effective for the achievement of the teaching goals?

Improving one's own teaching
Improve the weaknesses in the lesson and examine upcoming lessons.
Investigate a teaching plan for content that needs to be retaught.
Use the evaluation to reevaluate what kind of efforts the students made and how they changed, and modify the evaluation accordingly.

Improving the students' learning
Strive to enable each student to improve their abilities based on how well they achieved their goals.
Think about the students who reported different evaluation results from the teacher and try to figure out why the evaluations were different.

not know what they are supposed to be doing. Thus, when a teacher realizes that a student is not achieving his or her goals, it is important that he/she thoroughly investigates why this is happening, make the student aware of the problem, provide instruction so that the student can quickly fix the problem, and enable the student to engage in meaningful learning. The teacher must clearly indicate his/her intent to help the student overcome his or her difficulties. If the problem seems too difficult for the student, the teacher can make learning easier by breaking the problem up into small steps or starting with a more simplified version. Simply saying, "Do your best" may be encouraging, but it does not amount to teaching. When informing a student of the results of their assessment, it is important that the student walks away from the encounter feeling inspired to tackle and overcome the challenges.

Editor's note: In Japanese, the same word "Hiyoka" is used for 'assessment' and 'evaluation' and enhanced the feedback. Where the author refers to measuring students' achievement or otherwise evaluating their performance, it would appear to be better to use the word 'assessment'. But in cases where the results of assessments are used by teachers to appraise the effectiveness of their teaching, the word 'evaluation' seems more appropriate. However, in the case of translated titles of official documents, such as those cited above, which use the term 'evaluation', it is not appropriate to change their wording. Readers should understand that the focus of these documents is on appraising student performance and enhancing achievement. As the author suggests, many of these assessments are formative in nature, being intended to provide information to students to improve their learning and to provide feedback to teachers on effectiveness of their teaching.

Table 1

		Students' Self – Assessment	
		Achieved goal	Did not achieve goal
Students Evaluation by Teacher	Achieved goal	Tell student, "If you pay attention to these particular things, together we'll be able to work to even further improve your performance."	Some students may be too hard on themselves, so instruct them to confirm the goals of the lesson and have confidence as they work through their lessons.
	Did not achieve goal	Some students may be too easy on themselves, so reconfirm the evaluation standards and instruct the attitude toward learning that is to be achieved.	Determine and make the student aware of the reasons they did not achieve their goals, provide advice to help them improve their performance quickly, and provide guidance to help the student engage in meaningful learning.

Section 4.6: Textbooks and Teaching Guides

Takeshi Miyakawa

Under the Japanese school system, teachers must use textbooks that have been approved by the Ministry of Education, Culture, Sports, Science and Technology. There are six types of mathematics textbooks for elementary and junior high schools and about twenty types, by grade level, for high school. These are published by private publishing companies. The elementary school mathematics textbook, which excludes first grade mathematics, consists of about 100 B5-sized pages divided into two volumes. The junior and senior high school textbooks consist of 100 to 200 A5- or B6-sized pages. In compulsory education, elementary and junior high school level, textbooks for students are distributed for free by the national government. The numbers of pages were restricted by the government and then, they have been compiled to avoid any overlap in the curriculum across different grade levels and to ensure that students can learn all of the necessary content, and complete their practice exercises, in the number of hours allotted for the school year.

Textbook publishing companies also publish accompanying texts for teachers called teaching guides. Their format and structure varies somewhat by publisher, but they are usually comprised of a *practical* handbook, which explains textbook articles in red and teaching development methods, and a *theoretical* handbook of about the same size.

The practical handbook contains information on teaching systems, goals for each unit, and teaching plans, and provides highly detailed information about the teaching process, such as questions to ask students and their probable responses, and other teaching essentials.

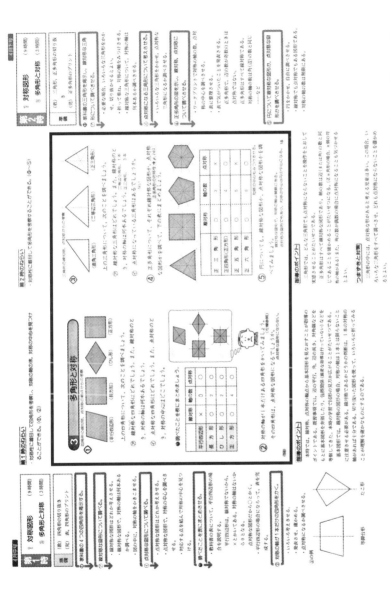

Figure 1: Pages from a sixth grade mathematics teaching guide published by Keirinkan. It provides explanations and answers to the textbook pages shown in a rubric in the frame in the middle of the pages, and shows the lesson sequence, teaching tips, and supplementary questions around the periphery.

The problem-solving approach characteristic of Japanese classes is derived from Lesson Study based on both theoretical and practical components. Figure 1 shows pages from an actual teaching guide *practical* handbook.

Lesson Study begins with materials research, and the meanings and significance of the specialized terminology and teaching materials used in mathematics education, which are necessary for this process, are explained in the *theoretical* handbook. This specialized terminology consists of terms specific to the field of mathematical education, as distinct from specialized mathematical terms or pedagogical terms. Take, for example, the terms "partitive division" and "measurement division," which are used to classify types of division problems (see Figure 2).

Editors' Note: Workbooks and Journals

Teachers usually use a workbook of exercises for students additionally.

There are a number of teachers' journals of mathematical education. They may be categorized in two. First type of journals is aimed to share the way of interesting teaching including subject matter and worksheet for children in a classroom. Second type of journals is focused in more developmental and challenging research with the terminology in teacher's guides and includes the reconstructed protocol of teaching process with major teacher's questioning, children's multiple answers and discussion.

Those Workbooks and Journals are used for sharing the experience of Lesson Study as well as teachers' guides.

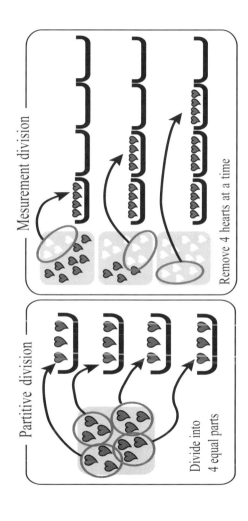

Figure 2: This image explains the terms "partitive division" and "measurement division" using the division problem 12÷4. "Partitive division" refers to dividing a whole into several equal parts, while "measurement division" refers to dividing a whole into groups of a certain number of elements.

Section 4.7: What Kinds of Teaching Materials and Aids are Used in Japan?

Hiroko Tsuji

Japanese classrooms utilize creative teaching materials and aids to enable students to experience the benefits and fun of mathematical ways of thinking and to cultivate a strong sense of quantities and figures in students.

In mathematics classes in the lower elementary school years, each child has their own mathematics activity set which the teacher uses regularly in the everyday course of teaching (Figure 1). Students use aids like geoboards and pattern blocks in elementary school classrooms, and teachers develop creative lessons that utilize these items. These aids are systematically chosen and used based on the nature of the activity.

Because of the need for ICT education, however, the national government is promoting the digitization of textbooks, teaching materials, and teaching aids by the textbook companies. Universities and boards of education promote the development of collection of links that support children's learning, and the national government has established the National Information Center for Educational Resources (Figure 2). This information is used in individual classrooms through video presentations. While traditional calculation devices, such as the abacus, are taught, students are also learning to use calculators and computers (Figure 3). Some schools are even promoting cooperative learning through groupware.

Mathematics activity sets

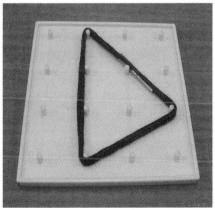

Geoboard

Figure 1

Rather than just looking at teaching aids as tools for imparting knowledge, teachers have to develop mathematics lessons designed to create a certain type of learning environment. They need to cultivate an environment that encourages students to learn and think on their own, and to pursue their own interests and their desire to know why and how things work.

In Japanese schools, students use workbooks in addition to textbooks. Workbooks are often used for the work students do outside of class hours. They help students absorb mathematical ways of thinking and expression, which they obtain through operational and experiential activities, and develop them into knowledge and skills.

Primary References
http://www.shinko-keirin.co.jp/
http://www.dainippon-tosho.co.jp/
http://www.nicer.go.jp/index_en.html

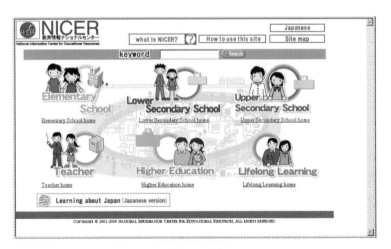

Figure 2

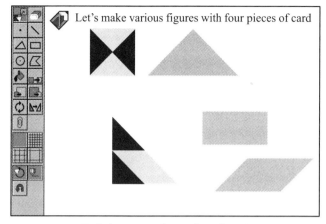

Dainippon Tosho: Educational
Mathematics software

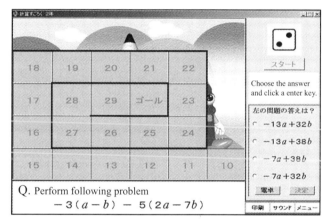

Keirinkan: Junior high school
educational Mathematics
software, Masunabi Do!

Figure 3: Sample efforts of the textbook companies

Section 4.8: What do Teachers and Teacher Trainees Think About Lesson Study?

Tadayuki Kishimoto

1. Teacher Trainee Attitudes toward Lesson Study

Under Japan's teacher training program in four year college (Table 1), prospective teachers must participate in a four-week period of teaching practice to become licensed to teach at the elementary and junior high school levels, and a two-week period of teaching practice to become licensed at the senior high school level. In addition to the actual practicum, they must receive preparatory instructions before the practicum and follow-up instructions afterwards.

Table 2 shows the results of a survey conducted by Nukui and Hirose (1997). This study examined the changes in attitudes held by 64 teacher trainees before and after they participated in their practicum. Before the practicum, teacher trainees reported a desire to have a classroom in which they would respect the independence of their students and prepare learning materials well suited to the actual needs of the students.

After the practicum, however, the teacher trainees became aware of the differences in the lesson plans they had prepared and the way the actual classroom experience unfolded, and they experienced difficulties in trying to communicate with their students.

Table 1: Necessary Credits of teacher's certification for bachelor degree in undergraduate program.

	Elementary	Middle	High
A) Academic Subjects	8	20	20
B) Subjects for Teachers	41	31	26
C) Subjects of A or B	10	8	16
D) Other Subjects	8	8	8

A) Special Academic Area. B) Including teaching methodology. D) Including Study of the Constitution

Table 2: Changes in the Attitudes of Teacher Trainees Before and After a Practicum (Nakai and Hirose, 1997)

Attitudes Before the Period of Teaching Practice

- I want a classroom where students can freely express their opinions.
- I want a classroom where the teaching materials are prepared based on an understanding of the students' needs.
- I want a classroom where children can learn independently.
- I want to incorporate students' ideas in the course of class.
- I want a classroom in which people can listen carefully to one another.

Attitudes After the Period of Teaching Practice

- I was able to develop the lesson while having fun communicating with the students.
- I tried to understand the individual personality of each student based on their everyday behaviors, and tried to communicate with them.
- I didn't know what to do when the children reacted differently than I'd expected.
- I realized that there are a lot of gaps between what the students want and what teachers want.
- There were major differences between what I planned and what actually happened in the classroom.
- Even if I understand a subject, it is difficult to study it along with the students and make them understand.

2. Teacher Attitudes toward Lesson Study

Table 3 shows the results of a survey conducted by Elementary School Section of the Research Department (2001) of Japan Society of Mathematical Education among 476 elementary school teachers in 2000.

According to this study, Japanese math teachers tend to:

(1) strive to facilitate communication between themselves and their students,

(2) incorporate specific hands-on and experience-based activities, such as experimental measurements,

(3) try to improve students' problem-solving capabilities by presenting problems that can be solved in a variety of ways,

(4) strive to impart theoretical understanding while simultaneously conducting skills practice,

and

(5) rarely conduct classes using an explanation-driven model in which the teacher gives an explanation and the students then solve problems.

Table 3: Teacher Attitudes Toward Lesson Study
(Elementary School Section: Research Department, 2001)

Topic	Always	Frequently	Occasionally	Never
My class emphasizes problem-solving according to the pattern: problem presentation→ independent work → development → summarization	11.9%	47.2%	37.1%	3.9%
I use problems that have multiple correct solutions.	1.1%	21.2%	52.6%	25.1%
I focus on having the students create problems rather than solving problems.	0.2%	11.5%	75.9%	12.4%
I respect the students' ideas and develop lessons based on my interactions with them.	17.3%	48.9%	32.0%	1.8%
I take students outside the classroom to gather materials or take measurements.	4.7%	32.4%	57.5%	5.4%
I emphasize tasks or creative activities that have mathematical content.	1.5%	25.4%	66.5%	6.6%
I use a textbook and plan lessons based on its lesson progression.	16.9%	40.1%	38.6%	4.4%
My class emphasizes skills practice and emphasizes the completion of workbook lessons.	4.2%	33.5%	55.3%	7.0%
I start out by explaining the important points and then later have students solve problems.	5.8%	28.4%	44.8%	21.0%

Comparisons of Features of Past International Comparative Studies

Section 5.1: Why Have Japanese Lessons Paid Attracted Attention and What Are its Features?

Hanako Senuma

1. Improving the Curriculum and Teaching Methods to Improve Academic Abilities

Since the Second International Mathematics Study (SIMS) was conducted in 1981 (Table 1), the US has been implementing curriculum revisions for the purpose of raising the level of academic abilities. Japan was the top of 20 countries in the SIMS seventh grade. The main reason that Japan obtained such a high score was accredited to the high level of its mathematics curriculum. But it is the teachers who are actually responsible for improving the curriculum. Thus, a video study on "A Comparison of Mathematics Instruction in Germany, Japan, and the US" was conducted on eighth grade as an optional study on the 1995 Third International Mathematics and Science Study (TIMSS 1995). This showed that Japanese mathematics lessons emphasize the process of problem solving, and since then several other countries have been hoping to improve academic abilities by conducting Japanese style lessons. However, it has already been 10 years since those classroom videos were taken, and today in Japan many new teaching methods, such as separating classes by level of proficiency, are being implemented.

Table 1: A focus on Japanese lessons by comparison of international study

International Association for the Evaluation of Educational Achievement (IEA) Main Study	TIMSS Video Study (IEA Option)	Year	Major factors that attention paid to Japan
Second International Mathematics Study (SIMS)		1981	
		1983	*A Nation at Risk* (Points out the low level of mathematics and science skills in the USA) Japan was the top of 20 countries in the eighth grade
		1986	Japan - USA seminar on Mathematical Problem Solving (A comparison of problem solving in the USA and Japan becomes the opportunity to introduce Japanese lesson in the USA. *The Open – Ended Approach*, a book on Japanese teaching methods, is translated into English in 1993.)
		1987	*The Underachieving Curriculum* (USA). (Argues that Japan's high scores are due to excellence of curriculum.)
		1989	*Curriculum and Evaluation Standards for School Mathematics* (USA). (Suggests that establishing a common curriculum nationwide will improve mathematics skills.)
		1991	*Professional Standards for Teaching Mathematics* (USA) (Training of high quality teachers is important for raising academic skills.)
Third International Mathematics and Science Study (TIMSS 1995)	Comparison of mathematics lessons in three countries	1995	
Third International Mathematics and Science Study – Report (TIMSS 1999)	Comparison of mathematics lessons in seven countries	1999	Results from mathematics lessons in three countries (TIMSS 1995 video study) increased interest for Japanese Lesson Study. (Japanese lessons focus on problem solving, which is linked to higher scores.)
Trends International Mathematics and Science Study (TIMSS 2003)		2003	Results are released on mathematics lessons in seven countries. (Show that Hong Kong and Japan, are both high scored countries and employ different teaching styles.)

2. Pose Problems to Raise the Level of Mathematical Ideas and Switching Lessons Between Whole Classroom Work and Individual Work

In 1999, eighth grade lessons in seven countries were videotaped as an optional study component of the TIMSS 1999 and the results were released in 2003. They indicated that Japanese teachers control their lessons well; the goals and summary statements are presented, teachers pose problems that require students to think (Figure 1), problems are presented that improve the students' abilities to make connections between ideas (Figure 2), alternative methods of solution for problem are examined, and teachers switch between whole classroom work and individual work as appropriate. The report showed that Hong Kong SAR and Japan had similarly high scores, but that they employ different teaching methods, thereby indicating that there is no single teaching method for improving academic abilities.

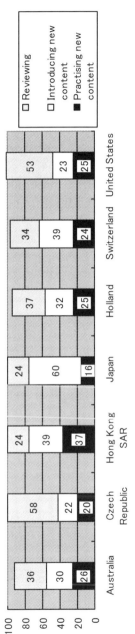

Figure 1: Japanese lessons emphasized introducing of new content

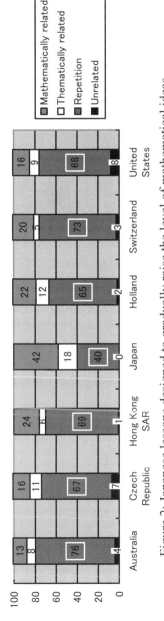

Figure 2: Japanese lessons designed to gradually raise the level of mathematical ideas

Recommended Websites: http://www.mext.go.jp/a_menu/shotou/shingi/index.htm, http://nces.ed.gov/timss/video.asp, http://timss.bc.edu/

Understanding Japanese Mathematics Lessons

Section 6.1: How do Japanese Teachers Explain and Structuralize Their Lessons?

Yoshinori Shimizu

Japanese lessons as "structured problem solving"

The following sequence of five activities has been described as the Japanese lesson pattern: reviewing the previous lesson; presenting the problems for the day; students working individually or in groups; discussing solution methods; and, highlighting and summarizing the main point (Table 1).

Table 1: The Japanese Lesson Pattern (Stigler & Hiebert, 1999, pp.79-80)

Reviewing the previous lesson
Presenting the problems for the day
Students working individually or in groups
Discussing solution methods
Highlighting and summarizing the main point

Teacher's Roles During the Lessons: Some Pedagogical Terms Shared by Teachers

Japanese teachers play several roles at each stage of their lessons. The following pedagogical terms are commonly used to describe such teachers' key roles.

"Hatsumon" at the presentation of a problem

"Hatsumon" means asking a key question for provoking students' thinking at a particular point in a lesson. At the beginning of the lesson, the teacher may ask a question for probing or promoting students' understanding of the problem. In a whole-class discussion, on the other hand, he or she may ask, for example, about the connections among the proposed approaches to the problem or the efficiency and applicability of each approach.

"Kikan-shido" during problem solving by students

"Kikan-shido", which means an "instruction at students' desk", includes a purposeful scanning by the teacher of students' problem solving on their own. The teacher moves about the classroom, monitoring students' activities mostly silently, doing two important activities that are closely tied to the whole-class discussion that will follow. First, he or she assesses the progress of students' problem solving. In some cases, the teacher suggests a direction for students to follow or gives hints to the students for approaching the problem. Second, he or she will make a mental note of several students who made the expected approaches or other important approaches to the problem. They will be asked to present their solutions later. Thus, in this period of the purposeful scanning, the teacher consider questions like "Which solution methods should I have students present first?", or "How can I direct the discussion towards an integration of students' ideas?" Some of the answers to such questions are prepared in the planning phase but some are not.

"Neriage" in a whole-class discussion

There is a term for describing the dynamic and collaborative nature of a whole-class discussion in the lesson. The term "Neriage" in Japanese refers to "kneading up" or "polishing up". In the context of teaching, the term works as a metaphor for the process of "polishing" students' ideas and getting an integrated mathematical idea through a whole-class discussion. Japanese teachers regard "Neriage" as critical for the success or failure of the entire lessons.

Based on his or her observations during "Kikan-shido", the teacher carefully calls on students, asking them to present their methods of solving the problem on the chalkboard, selecting the students in a particular order. The order is quite important to the teacher both for encouraging those students who found naive methods and for showing students' ideas in relation to the mathematical connections that will be discussed later. In some cases, even an incorrect method or error may be presented, if the teacher thinks it would be beneficial for the class. Students' ideas are presented on the chalkboard, to be compared with each other with oral explanations. The teacher's role is not to point out the best solution but to guide the discussion by the students towards an integrated idea.

"Matome" as summing up

"Matome" in Japanese means "summing up". Japanese teachers think that this stage is indispensable to any successful lesson. It is identified as a critical difference between the U.S. and Japanese classroom activities (Fujii, et al., 1998). According to the U.S.-Japan comparative analysis, at the Matome stage Japanese teachers tend to make a final and careful comment on students' work in terms of mathematical sophistication.

Generally speaking, in the Matome stage what students have discussed in the whole-class discussion is reviewed briefly and what they have learned through the lesson is summarized by the teacher.

Some practical ideas shared by Japanese teachers.
Ensuring the Student's "Ownership"

During the discussion, each solution method is labeled with the name of student who originally presented it. That is, the name of student who presented the solution will be written, or a small magnet card with his/her name will be put on the chalkboard. Thus, each solution method is referred using the name of student in the discussion. This practical technique may seem to be trivial but it is very important to ensure each student's "ownership" of presented methods.

"Bansho": Effective Use of Chalkboard

Another important technique used the teacher relates to the use of chalkboard, which is referred as "Bansho" by Japanese teachers. Teachers usually try to keep all that is written during the lesson on the chalkboard without erasing if possible. From the learner's perspective, it is easier to compare multiple solution methods if they appear on the chalkboard simultaneously. Also, the chalkboard can be a written record of the entire lesson, which gives both the students and teacher a birds-eye view of what has happened in the class at the end of each lesson.

Teaching and Evaluation as Two Faces of the Same Coin

Teachers are conducting formative evaluations during their lessons to obtain instantaneous feedback on their instructional techniques. Such evaluations are embedded in each teacher's role described above.

When the teacher moves about the classroom for "Kikan-shido" during problem solving by students, he or she is monitoring students' activities silently to assess their status or making suggestions to individual students who need help or guidance. Thus, it is important to see integrating teaching and its evaluation as two faces of the same coin.

The Importance of Evaluation as Incorporated with Teaching

- Teaching and evaluation activities are done to ensure that the teaching goals established based on the curriculum and teaching plans are being achieved by the students with whom the teacher is currently working.
- Teacher evaluations are intended to enhance teaching practices. For example, they can allow teachers to ascertain the effectiveness of their teaching practices and help them improve their teaching plans by incorporating the results into their teaching.
- For students, evaluations are an important tool for making them aware of how well they are learning, giving them an opportunity to adjust their behaviors, and enabling them to set their own learning goals.
- Incorporating teaching and evaluation in the teaching process makes it possible to plan comprehensive evaluations that look at both the teaching process and its results.

Section 6.2: How do Japanese Teachers Evaluate Their Students in Their Lessons?

Hiroyuki Ninomiya

Teachers are conducting formative evaluations during their lessons to obtain instantaneous feedback on their instruction techniques. These evaluations may be completed by students individually or in groups. Let us remind at first general image of the Japanese classroom. To get the general idea, we gather together information already explained in the text, add some comments and give actual examples.

Image of the Japanese Classroom (*The Teaching Gap*, Stigler & Hiebert, 1999)

1. Teacher reviews previous lesson and assigns a problem that was not finished.
2. Students present solution methods they have found, and teacher summarizes.
3. Teacher presents task for the day and asks students *to work on it independently* (task is to invent problem for classmates to solve).
4. Teacher instructs students to work in small groups. Leaders of groups share problems with teacher, who writes them on board. Students copy problems and begin working on them.
5. Teacher highlights a good method for solving these problems.

The Objectives of Desk Instruction

"Desk Instruction" occurs when a teacher walks around between student desks while students are working individually to examine students' learning and to provide help or guidance as needed. Desk instruction has two objectives. The first is to ascertain how well individual students are learning (their level of recognition) so as to create more active argumentative communication and deepen group thinking during the presentation period after the individual study period, and to formulate discussion ideas so that a diversity of opinions can be expressed during the group learning period. The second objective is to help eliminate individual errors and to improve students' academic abilities. Therefore, following questions should be considered to make clear the objectives of desk instruction (Table 1).

Table 1: Are the objectives of desk instructions clear to you?

• To know the level of students' understanding?
• To ascertain types of reactions (as preparation to choose students who will present information)?
• To support students who have trouble learning?
• To support students' group activities?

Integration of Teaching and Evaluation

Teaching and evaluation activities are done to ensure that the teaching goals established based on the curriculum and teaching plans are being achieved by the students with whom the teacher is currently working.

For teacher, evaluations are intended to enhance teaching practices. For example, they can allow teachers to ascertain the effectiveness of their teaching practices and help them improve their teaching plans by incorporating the results into their teaching.

For students, evaluations are an important tool for making them aware of how well they are learning, giving them an opportunity to adjust their behaviors, and enabling them to set their own learning goals.

Incorporating teaching and evaluation in the teaching process makes it possible to plan comprehensive evaluations that look at both the teaching process and its results. Hence, two senses of the integration of teaching and evaluation are used as in what follows.

Two Senses of the Integration of Teaching and Evaluation

1. Integration in the sense of using the evaluation results in future teaching development efforts and teaching plans: Evaluations should not be performed at the end of a teaching activity, but during the activity. This way the teacher can use the results to examine and make adjustments to the teaching practices they have used thus far, and can either adopt new or supplementary teaching practices. There should be an emphasis on formative evaluations.

2. Integration in the sense of using the evaluation process itself as a teaching tool: Evaluations serve as a means of teaching students. For example, an evaluation that tells the student they "worked really hard" simultaneously helps stimulate the student's desire to learn. Below are some examples of a range of rich evaluative comments that Japanese teachers use to encourage students, to provide constructive feedback to assist learning, and to foster more effective participation in mathematics classrooms (Table 2).

Table 2: Some examples of evaluative feedback between teachers and
students

Verbal Assessments which value students' endeavor	You have improved a lot, as you always try to think deeply. You have concentrated a lot. Wonderful! Great! You could solve such many problems beautifully.
Verbal Assessments which make students try harder	It's a pity, but it is OK. You can do it. Try better next time. You can do it well if you apply yesterday's outcomes. You had better try another angle. Be confident!
Verbal Assessments which stimulate students' interest and motivation	You have been working with a lot of confidence. Now it seems you like to learn about fractions. Yours is such a good question. It interests everyone.
Verbal Assessments which value students' ability	Since you have understood his idea so well, please explain it to everyone. Wow, you are the champion of multiplication! Your explanation is very clear and really helpful to understand.
Verbal Assessments which gives energy and hope for learning	You seemed to think you could not understand today's problem. Ok, I will work with you tomorrow until you can be satisfied with your understanding. You had made a lot of careless mistakes because you were in a rush, but now you have very few. You are thinking deeply and more carefully.
Verbal Assessments which value students' contributions	Because of your question, we got some good hints for solving this problem. Because you carefully explained your idea, many people could understand quite a lot.

CHAPTER 2

Methods and Types of Study Lessons

Preparation for Lessons

Section 1.1: Annual Teaching Plan as a Plan for Nurturing Students

How should Annual Teaching Plans be Created to Impart Useful Skills and Creative Ways of Thinking?

Yasuhiro Hosomizu

Annual Teaching Plans are created by each school based on the National Course of Study. However, it is important to develop the Annual Teaching Plans into plans that are suitable for the children being taught. The following are several important points that are regarded on the creation of Annual Teaching Plans designed to "impart useful skills and creative ways of thinking."

1. Connections should be "pasted" rather than "taped"

Think about when moving on to different grade levels or the next subject units as having to be connected by paste (which requires overlapping) rather than tape (which can be done side-by-side) or "end-on" where little or no time is spent making connections and reviewing between lesson, or between topics, or within a grade level, or between grade levels. [Editors' note: Japanese textbooks are written based on what is already learned, there is little or no overlapping and review. Textbooks are small compared, for example, with a typical textbook in the U.S.A.] That is, we need to impart useful skills and creative ways of thinking by repetition and reviewing and building on what they have already learned.

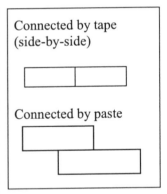

At first glance, an overlapping approach appears to be extremely wasteful. However, because it is easier to make subject unit creation that suit the students' actual needs, it makes it easier for them to learn what is being taught, and given the differences in individual learning among students in group-based lessons, this is an appropriate way to develop effective teaching. For example, if

Figure 1: Annual Teaching Plan as a Plan for Nurturing Students

Annual Teaching Plan (sample) which analyzes the teaching curriculum by the fundamental/basic pillars in each area, divided into lower (1st and 2nd grades), middle (3rd and 4th grades), and higher (5th and 6th grades) grade levels.

		Grades 1 & 2	Grades 3 & 4	Grades 5 & 6
Mathematical concepts and expressions	Integers	• One-to-one correspondence • Collective numerals and ordinal numbers • 1-digit to 4-digit numbers (up to 10,000 position) • Relative size of tens and hundreds • Counting in groups of numbers. Divide into equal parts	• Numbers up to 1 trillion • Structure of numbers (size of a factor of 10, a factor of 100, 1/10, 1/100) • Relative size of numbers • Odd numbers, even numbers • Approximation, rounding	• Types of division • Divisors, multiples, common denominators, common multipliers
	Decimals	*See content handling (1) of the National Course of Study.	• Meaning and structure of decimals (down to the third decimal place) • Relative size of numbers • Approximation	• Structure of the decimal notation system
	Fractions	• Fractions (1/2, 1/4, etc.)	• Meaning and expression of fractions • Equivalent fractions • Simple size comparisons of fractions with unlike denominators	• Reduction • Equivalence of fractions • Quotient fractions
Calculations	Addition	• Meaning of addition • Addition of 1-digit numbers • Addition of 2- and 3-digit numbers • Longhand addition	• Addition of decimals • Addition of fractions with a common denominator	• Addition of fractions with unlike denominators • Mixed calculations with decimals, fractions, and integers
	Subtractions	• Meaning of subtraction • Subtraction of 1-digit numbers • Subtraction of 2- and 3-digit numbers • Longhand subtraction	• Subtraction of decimals • Subtraction of fractions with a common denominator	• Subtraction of fractions with unlike denominators • Mixed calculations with decimals, fractions, and integers
	Multiplication	• Meaning of multiplication • Multiplication table	• (2- and 3-digit numbers) × (1- and 2 digit numbers) • Multiplying three digit numbers • Longhand multiplication • Decimals × integers • Fractions × integers	• (Integers, fractions) × fractions • Meaning of the multiplication of decimals • (Integers, decimals) × decimals • Mixed calculations with decimals, fractions, and integers
	Division	*See content handling (2) of the National Course of Study.	• Meaning of division • Meaning of the remainder • Calculations when the divisor is a 1- or 2- digit number • Relationship between dividend, divisor, quotient, and reminder • Decimals ÷ integers	• Fractions ÷ integers • (Integers, fractions) ÷ fractions • Meaning of the division of decimals • (Integers, decimals) ÷ decimals • Mixed calculations with decimals, fractions, and integers
Calculation properties and rules		• Correlation between addition and subtraction • Properties and rules of multiplication (multiplicative identity property, commutative property)	• Commutative, associative, distributive properties of multiplication • Properties of division • Formulas that mix the four operations, the meaning and order of calculations in formulas using brackets ( )	• Properties and uses of calculations *See content handling (3) of the National Course of Study.
Sense of number		• Understanding numbers as the product or quotient of other numbers such as the sum, difference or multiplication • Approximate numbers • Approximate calculations *See content handling (4)	• Understanding numbers as the product or quotient of other numbers (related to divisors, multiples)	

we think about children's developmental stages, the systematization of teaching content, and teaching methods suited to the students' needs, we can try to think of the teaching content in three stages of lower (first and second grades), middle (third and fourth grades) and higher (fifth and sixth grades) grades (Figure 1).

The National Course of Study stipulates that division should be taught by teaching division that involves count back in third grade and division with single-digit and double-digit divisors in fourth grade. However, for example when third graders are studying by using cards, searching for a card with the same quotient, simple division problems often emerge even though the students are not particularly familiar with the count back like $30 \div 3$ or $33 \div 3$. If these calculations are addressed during this lesson, it makes division learning in fourth grade much easier.

By addressing curricular content in multiple grades, the students can learn through repetition in both years. By applying more weight to the topic in one of these years, the lesson is not just taught repetitively, but in such a way as to enable students to learn in a progressive or developmental manner. In addition, by addressing curricular content in multiple grades, teachers can choose to emphasize different elements of the lesson in different years.

2. Divide subject units into small units for repetitive learning

Under the paste-connected (overlapping) approach, repetitive learning can also be implemented in a single grade level. For example, the 16 hours of lessons on division are separated into three lessons in Plan A and five lessons in Plan B (Figure 2). Instead of a format in which division lessons are done all at once, and then never appear again once they are completed, educators ensure that division lessons appear repeatedly so that students can learn through repetition. If teachers can use the content from a previous lesson and expand it, this will improve the students' ability to think developmentally and to experience the excitement of expanding their knowledge. Teaching plans that repeat and build on the basics and fundamentals are also important in terms of utilizing evaluations in the teaching process.

Month	Unit
April	1. Large numbers 　1. Large numbers 　2. Structures of whole numbers
May	2. Circles and spheres 　1. Circles 　2. Spheres
June	3. Division 　1. Longhand division 　2. Division of tens and hundreds 　3. The principles of division 4. Dividing by a single-digit number 　1. Division with a double-digit quotient 　2. Three digits ÷ one digit 　3. What kind of formula will it be?
July	5. Organization of materials
September/October	8. Dividing by a double-digit number 　1. Double-digit division 1) 　2. Double-digit division 2) 　3. The principles of division

Plan A: 4-Year Teaching Plan (Sample)

Month	Unit
April	1. Large numbers 2. Structures of whole numbers 1. Longhand division
May	1. Structures of whole numbers
June	1. Circles 2. Spheres 1. Division of tens and hundreds 2. The principles of division
July	1. Organization of tables 2. Organization of materials 1. Division with a double-digit quotient 2. Three digits ÷ one digit 3. What kind of formula will it be?
September/October	1. Double-digit division 1) 2. Double-digit division 2) 1. Angles 1. The principles of division

Plan B: 4-Year Teaching Plan (Sample)

Figure 2: Annual Teaching Plan (Plan B) created to allow teachers to conduct review lessons while expanding the curriculum.

Section 1.2: Teaching Plans in Which Questions Continuously Emerge
How to Develop Lessons in Which Students Say, "Wow, It is Really Easy to Calculate This Way!"

Yasuhiro Hosomizu

In the process of thinking about how to approach lessons in Long-Hand Multiplication (in third grade), we will think about how to develop lessons so that students learn the long-hand method while enjoying the process of thinking through reasoning.

1. Create opportunities where students can experience the excitement of thinking in mathematical terms

In this subject unit, the students will learn the long-hand method of multiplying two double-digit numbers. Once students learn the long-hand method, lessons often tend to become very dry and boring, designed only to enable students to master the skill and quickly and accurately compute problems, like a machine. Unfortunately, this can cause students who may have started to find it "fun to think" and who may have taken the initiative in learning things in the past to become passive.

Therefore, a situation should be created in which this skill can be mastered while new problems emerge. That is, we will think of a way to develop the lesson so that students will naturally start to notice the way of long-hand multiplication in the process of solving the problems. When they do this, they will start to realize that some patterns seem to be emerging. However, they will not be able to see the patterns clearly. At this point, the students will take steps to try to obtain new information. Teachers must pay attention to these steps.

This series of activities can cultivate the students' ability to look for and construct patterns through inductive thinking. Once they can see a pattern, new questions arise: "Why is this true?" "Does

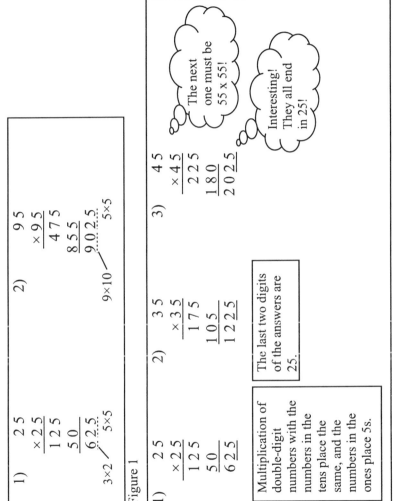

Figure 1

Figure 2

the pattern always work?" They will then take action based on this new awareness of the problem. In this way, the teacher can improve the ability of students to think developmentally and deductively.

This example uses the multiplication of two numbers with the same number for the tens digit and 5 for the ones digit. The patterns that appear in these kinds of calculations are shown in Figure 1. We want students to learn the joy of discovering a pattern that makes calculation easier or the activity of working together to find patterns and think about why they occur.

2. Developing creative mathematical activities

Here we have emphasized the importance of creating as many opportunities as possible in which the students can take the initiative in their learning and for teachers to watch the process. For this to happen, the teacher has to devise the lesson so that the following types of mathematical activities can emerge.

2.1. Mathematical activities to make students realize that patterns might exist

Line up the double-digit multiplication problems on the chalkboard as shown in Figure 2, and ask the whole class to work out the answers together. This provides the opportunity for students to start to realize that patterns might exist.

The students are just completing problems that have been given to them, but the teacher has to watch their reactions as they do this. As they perform the calculations, the students will start to mutter things like, "The next one has to be 55 x 55!" and "Interesting! They all end in 25!" Again, they are reading ahead and trying to move forward. The teacher should encourage those murmurings and movements.

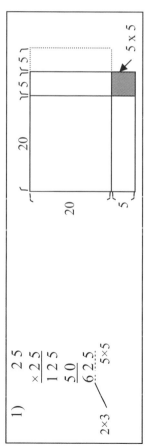

1)
$$\begin{array}{r} 2\,5 \\ \times\,2\,5 \\ \hline 1\,2\,5 \\ 5\,0 \\ \hline 6\,2\,5 \end{array}$$

5×5

2×3

20   20   ⌜5 ⌝⌜5 ⌝   5 × 5

Figure 3

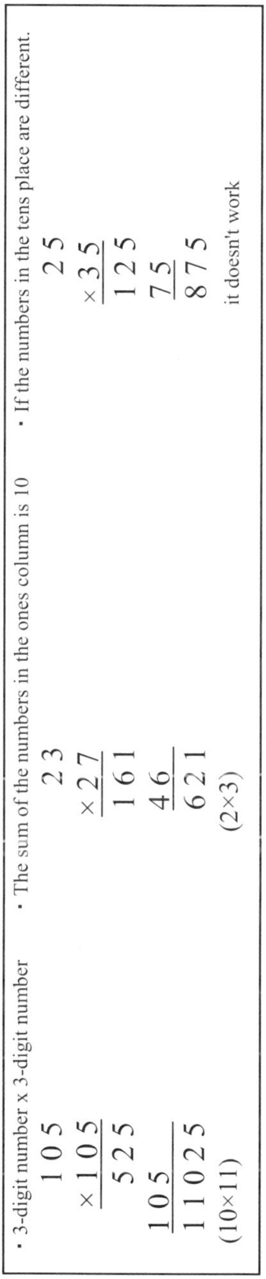

· 3-digit number x 3-digit number

$$\begin{array}{r} 1\,0\,5 \\ \times\,1\,0\,5 \\ \hline 5\,2\,5 \\ 1\,0\,5 \\ \hline 1\,1\,0\,2\,5 \end{array}$$

(10×11)

· The sum of the numbers in the ones column is 10

$$\begin{array}{r} 2\,3 \\ \times\,2\,7 \\ \hline 1\,6\,1 \\ 4\,6 \\ \hline 6\,2\,1 \end{array}$$

(2×3)

· If the numbers in the tens place are different.

$$\begin{array}{r} 2\,5 \\ \times\,3\,5 \\ \hline 1\,2\,5 \\ 7\,5 \\ \hline 8\,7\,5 \end{array}$$

it doesn't work

Figure 4

Once they see the characteristics of the numbers lined up, they will start to notice patterns. "These are all multiplying identical double-digit numbers when the numbers in the tens place are the same and the numbers in the ones place are 5." And, "The last two digits of the answers are 25."

2.2. Mathematical activities for obtaining information that will help them recognize the patterns

After they solve 45 x 45, ask "Now can you quickly figure out the solution to 95 x 95?" Once you say this, the students will begin working on their own to obtain information that will help them see the patterns.
- 55 x 55 = 3025, 65 x 65 = 4225... They will start to collect information by doing the problems in order.
- They will review the information in the problems leading up to 45 x 45.
- They will try to solve the simple case of 15 x 15.
- After calculating 95 x 95 = 9025, they will look for all kinds of ways to try to find the patterns.

Pay attention to the students' reaction. The problem is not just to see if the students figure out the patterns, but to pay attention to the efforts they make to try to find them and how they go about doing this. By telling them what some of their classmates are doing, you can give clues on how to make approaches for the rest of the class and improve their ability to take action. Then they will notice that the first two numbers of the response = (number in the tens place) x (number in the tens place + 1).

2.3. Mathematical activities to encourage students to think about the reason why the patterns apply

In a lesson where students discover patterns, it is not enough to end the lesson after the patterns are found. We want to encourage

Students' Learning Activity	Teaching Tips
1. Perform the calculations shown. $\begin{array}{r} 25 \\ \times 25 \\ \hline 125 \\ 50 \\ \hline 625 \end{array}$　$\begin{array}{r} 35 \\ \times 35 \\ \hline 175 \\ 105 \\ \hline 1225 \end{array}$　$\begin{array}{r} 45 \\ \times 45 \\ \hline 225 \\ 180 \\ \hline 2025 \end{array}$ Multiplication of identical double-digit numbers when the numbers is in the tens place are the same and the numbers in the ones place are 5s. 2. Look for the patterns ⎢ What is the result of 95 x 95? ⎢ Obtain the information needed for finding the patterns. $\begin{array}{r} 15 \\ \times 15 \\ \hline 75 \\ 15 \\ \hline 225 \end{array}$ ↓ $\begin{array}{r} 25 \\ \times 25 \\ \hline 125 \\ 50 \\ \hline 625 \end{array}$ $\begin{array}{r} 35 \\ \times 35 \\ \hline 175 \\ 105 \\ \hline 1225 \end{array}$ $\begin{array}{r} 45 \\ \times 45 \\ \hline 225 \\ 180 \\ \hline 2025 \end{array}$ ↑ $\begin{array}{r} 95 \\ \times 95 \\ \hline 475 \\ 855 \\ \hline 9025 \end{array}$ Find the patterns	Write the problems at left along with the answers up on the chalkboard for everyone to examine. Pick out the mumblings or actions of students who start to notice the patterns, and make notes on the chalkboard. Just as the students realize there's a pattern, have them solve the problem shown at left. Make students keep track of patterns identified in a notebook

Figure 5 (Continued to page 85)

children to always think about questions like, "Why does the pattern works" and "How far can it be taken?" Thus, we want to create opportunities for the students to think about the reason why the pattern works. Given that this lesson is for third grade, however, it may be difficult for some students to come up with reasonings. In this case, the teacher will have to provide support, encouraging students to check whether the patterns found will apply in other situations and to think about the patterns using diagrams.

2.4. Mathematical activities using a developmental approach

The students can probably understand that the patterns work in all two-digit problems (up to 95 x 95), so try to plan activities using a developmental approach, like expanding the lesson to triple-digit numbers or changing the conditions, as shown in Figure 3. Figure 4 shows a plan of further developing the lesson.

- Teaching Plan for this Class
1. Goal
   To discover patterns while doing interesting computations and to enable students to think about the framework or range of numbers in which those patterns apply.
2. Lesson Plan (Figure 5)

The last two digits of the answer are 25.

The answers increase in increments of 200.

The first two numbers of the response =

(number in the tens place) x (number in the tens place + 1)

| | Explain the steps taken to discover the patterns, and confirm that students understand that it's not just the answer, but the process that is important. |

3. Explain the patterns found

What kinds of patterns did you find?

Last two digits are 25
First two digits are
9 x 10

$$
\begin{array}{r}
95 \\
\times 95 \\
\hline
475 \\
855 \\
\hline
9025
\end{array}
$$

(number in the tens place) x (number in the tens place + 1)

4. Think about the reason why each pattern applies

Will the pattern work in every situation?

| | Explain the structure of the patterns in simple terms the children understand. |

15 x 15	25 x 25
Diagram is left out here	Diagram is left out here

| | If it is difficult for the students to come up with a framework in which a pattern applies, use diagrams. |

5. Think about the development of the problem

Will it still work as the numbers grow larger?

Will it still work if the numbers in the tens place are not the same?

| | Get students to think about how the patterns can be expanded. |

Figure 5 (Continued from page 83)

Section 1.3: Developing Creative Teaching Strategies Aimed at Imparting Diverse Ways of Thinking and Fostering Enjoyment of Learning

Kozo Tsubota

1. Purpose of the Lesson Plan

Figure 1 shows an example of the items that might be included in a lesson plan. These items are included for the following reasons:

A) *To explain the content of the teacher's plan to research lesson observers.*

Explanation of study topics stating the highlights of the lesson: Describes the teaching materials, students' perspectives, students' demeanor, and strategies for improving the students' demeanor.

B) *To position the goals of this class in terms of the annual lesson plan or National Course of Study.*

Unit lesson plan: Describes how students will be able to learn new material based on what they have already learned, and what kinds of questions will be used to achieve this.

C) *To enable teachers to develop lessons to facilitate achievement of goals.*

States the goals of the lesson, the topics to be covered, including tentative questions and possible answers, observations, judgments, and feedback criteria (evaluation plan) that the teacher will use based on students' responses, blackboard plans to ensure that all material covered is reviewed at the end of the lesson, and prepared materials such as teaching aids and handouts.

Subject of Lesson Plan: Name of Teacher:

I Unit (Topic)

II Unit (Topic) Lesson Plan

1 Goals

2 Teaching Materials

(Mathematical background, tie-ins with past lessons, teaching sequence)

3 Students' demeanor

4 Time allocation

(number of hours spent on the overall unit and on each component of the lesson)

III Lesson Goals

Identify what students are to achieve by clarifying the nature of the lesson based on the four perspectives described in the National Course of Study: (1) interest, eagerness, and attitude, (2) thought and judgment (3) skills and expression, and (4) knowledge and understanding.

IV Lesson Guidelines

Implement diverse independence-inspiring activities that encourage students to experience the fun of learning.

V Teaching/Evaluation Plan, Evaluation Standards

1 Evaluation standards for the unit

2 Plan for developing the lesson plan in each class hour and evaluation from different perspectives

VI Current Class Lesson (Class ___ out of a total of ___ classes)

1 Lesson goals

2 Preparation and materials

3 Detailed development plan

(goals and questions, strategies for helping students, evaluation and consideration criteria)

VII Current Class Evalation

Comprehensive evaluation from different perspectives based on the evaluation standards.

Figure 1: Information included in the lesson plan

2. Teaching Strategies Aimed at Imparting Diverse Ways of Thinking and the Fun of Learning

Because the goal is to ensure that students learn the content of the current lesson on their own based on what they already know, students are encouraged to propose ideas about what they need to know to solve the problem posed in the lesson, and the teacher creatively designs the lesson so that students learn about the topic through that discussion. In the materials research process prior to the lesson, the teacher prepares problems that allow students to express their own ideas, and problems that enable students to learn the target material based on what they already know. The teacher then tries to imagine what ideas the students might use to solve the problem, develops creative questions for eliciting solutions, and prepares questions that help students understand and appreciate the importance of the concepts being taught.

Explanation of lesson example

Specific goal: Impart the diversity of ideas and the fun of learning through an open-ended approach

Study topic: Lesson on solid figures learned by expressing different projected diagrams

1. Unit: Solid Figures

1) Unit Goals

To deepen students' understanding of basic solid figures through activities such as observing and constructing figures, and to enable them to understand and discuss the structural components of a figure and their relative position.

	Activity	Precautions

Introduction	(First lesson in two lesson hours)
	Yesterday, we studied pyramid through unfolding and designing. Today, we explore the cylinder. Let's draw a unfolded diagram of a cylinder (freehand)
	Response 1:Typical
	Projected diagram  

	Teacher's question:
	Does the unfolded diagram you drew in your notes look like this?
	
	Response 2:
	   Unfolded diagram
	   where the position of the circles differs
	Response 3:
	   Projected figure without a rectangle
	Teacher's question: What is this? What does this mean? 

	Summary on the ideas in the construction of a unfolded diagrams. 
		◆ Check that the circumference of circles and the length of the sides of the rectangle are the same, and that the circles intersect with one point on the rectangle.
		◆ Knowing how to draw the center of the circle on the rectangular edge
		◆ Even if the position of the circles is off, be sure that they appear to be the same.
		◆ Have one child draw the diagram on the chalkboard and another give the explanation.
		◆ Be sure that students have found creative ways to cut off the curves.
		◆ There is no edge on the side face of cylinder and can be cut any ways.
		◆ Identify and discuss creative strategies for dealing with both the upper and lower sides.

2) Teaching Materials

The teaching standards show a lesson structure that teaches solid figures based on lessons using two-dimensional figures. The textbook uses a typical projected diagram to address activities for expressing solid figures as two-dimensional figures and activities for making solid figures from a projected diagram. Activities that enhance students' understanding of the structural components of a figure and their relative positions by encouraging them to think about what they could do if the typical projected diagram were conducted to make the lesson more fun and to enhance learning. By viewing and comparing diverse projected diagrams, students are expected to learn from one another the diversity of ideas that are used for creating projected diagrams and the graphical beauty of various projected diagrams. They are also able to learn the conditions needed for constructing an object from a projected diagram, and the methods of deriving one projected diagram from another.

3) Unit Lesson Plan (omitted) (After the teaching order within the unit) Class 2 of 3: "Projected Diagram of a Cylinder"

4) Students' Demeanor (omitted)

2. Current Class Lesson: "Projected Diagram of a Cylinder"

1) Current Class Goals

To learn methods of constructing a cylinder from the projected diagram of a two-dimensional figure.

2) Lesson Guidelines

Students are already familiar with the projected diagram of a pyramid. Building on this experience, the cylinder lesson allows students to learn even more independently.

3) Lesson Plan

(for the implemented lesson, see www.criced.tsukuba.ac.jp/math/video)

	Activity	Precautions
Problem Posing	Based on the yesterday's activities about unfolding of a pyramid, let's explore the designs with a cylinder. Let's draw a unfolded diagram and cut it out, and develop your designs.	◆ Remind students of the yesterday's activity.
Problem Solving by students	Students draw a unfolded cylinder, cut it and fold. (Second lesson in two lesson hours)	◆ In the process, find students' failed experience and share difficulties behind problems that they may confront. ◆ Let them know about trial and error approach and not worry about error.
Sharing	Comparing developed various designs after cutting out cylinder.	◆ Let students share ideas among each other. ◆ Let them be aware of goodness and usefulness.
Summarizing	Cylinder is constructed with two circles and a curved face. There are many possible unfolded diagrams of cylinder. There is no edge beside of a cylinder and various ways of cutting a curved face. The case of rectangle is the most simple way of cutting.	◆ Comparing the findings with the experience of the pyramid, let them reflect and know the features of cylinder.

Unique Japanese Lesson Development – Models and Examples

Section 2.1: The Problem-Solving Oriented Teaching Methods and Examples

Satoshi Natsusaka

1. What is a problem-solving oriented lesson?

There are several phases to the process of solving a problem. "Problem-solving oriented lessons" are lessons that focus on those phases. This is the typical approach used in Japanese mathematics classes throughout Japan. The stages are established by each school or region, and classes are often structured around them.

There are slight differences in the phases established by different schools and regions, but lessons are usually comprised of four or five phases, shown at right, which include "identifying the problem," "developing a solution," "progression through discussion," and "summarizing."

This model was explained based on the work of Polya, Dewey and Wallas who studied the problem-solving process.

Polya's four phases: (1) Understand the problem, (2) devising a plan, (3) carrying out the plan, and (4) looking back.

Dewey's five phases: (1) Experience a difficulty, (2) define the difficulty, (3) generate a possible solution, (4) test the solution by reasoning, and (5) verify the solution.

Wallas's four phases: (1) Preparation, (2) incubation, (3) illumination, and (4) verification.

Problem-solving oriented lessons are often incorporated in research lessons, but adopting an approach that focuses on fitting children's thinking process as a group and making in into a "pattern" and partitioning the lesson time (usually 45 minutes) into several sections has been shown to pose several problems. These difficulties would come from asynchrony between teaching phase and children's thinking processes.

Identifying the problem   About 10 min.

Students and a teacher read the problem, understand the problem.

Students figure out what the problem is by reading and listening to the teacher's instructions and by discussing it amongst themselves.

Students check the similarities and differences between what they have already learned and what they are learning in the lesson.

Students develop a perspective on ways of solving the problem and the solution.

Developing a solution   About 15 min

Students think about the problem on their own and try to find the solution by themselves.

The teacher walks around the room providing instructions, offering hints to those who cannot come up with ways to solve the problem, and encouraging students who have already found the solution to think of alternative solutions.

Progression through discussion   About 10 min.

Three to five of the students who were able to get different answers explain their approaches to the rest of the class.

After listening to the explanations, the students reach a common understanding of better solutions by discussing the strong and weak points of each approach proposed and identifying what they have in common.

Summarizing   About 10 min.

Summing up the important points addressed during the lesson.

Challenge students with similar problems, strive to firmly implant the lessons learned.

"Sample Patterns for Establishing the Phases"

Understand – get a insight – investigate – confirm – sum up

Understand the problem – make predictions – pursue a solution – apply

Seek – think – create – review

Figure 1: Structural model of a problem-solving oriented lesson

2. Examples of a problem – solving oriented lesson

2.1   Unit: Division, 4th grade

2.2   Division in each grade level

Third Grade

Students first begin to learn division in third grade. At this level, they learn that there are two types of division: "partitive division" and "measurement division." They work with numbers within the range of the 9 × 9 multiplication table, solving a calculation like 12 ÷ 3 (that is a calculation where the divisor and quotient are both single digit numbers)

They learn that there are situations in which a number cannot be divided evenly, like in the calculation 13 ÷ 3, expanding their understanding of divisions and learn that there are division with remainders.

Fourth Grade

In fourth grade, students learn to divide two- or three-digit numbers by one- or two-digit numbers, like 96 ÷ 8 and 288 ÷ 12. The goal is to introduce the long-hand format and to enable students to understand the meaning of division and develop their ways to compute division problems. Even as the range of numbers they use expands, students are able to understand new computations based on the division problems they learned in the previous year.

Students also study the relationship between the dividend, divisor, quotient and remainder and study the properties of division.

Fifth and Sixth Grades

In fifth grade, students learn the significance of and processes for computing a decimal ÷ an integer, and a decimal ÷ a decimal.

In sixth grade, they learn the significance of and processes for computing a fraction ÷ an integer, and a fraction ÷ a fraction.

Table 1: Grades in which the various types of division are taught (elementary school)

	1st Grade	2nd Grade	3rd Grade	4th Grade	5th Grade	6th Grade
Integers			○	○		
Decimals					○	
Fractions						○

A partitive division problem

Suppose you divide 12 oranges equally between three people. How many does each person get?

A measurement division problem

Suppose you divide 12 oranges by giving three to each person. How many people will get oranges?

Long division

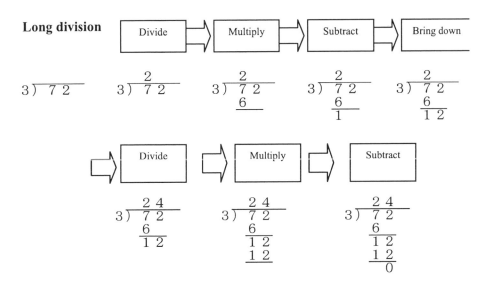

Relationship between the dividend, divisor, quotient, and remainder

　Dividend = divisor × quotient + remainder

Properties of division

If　　　　　　$a \div b = c$,

then　　　　$(a \times m) \div (b \times m) = c$

　　　　　　$(a \div m) \div (b \div m) = c$

2.3 About this lesson

(1) Goal

For students to think about how to compute problems in which the number in the tens place in the dividend cannot be evenly divided by the divisor, such as in 72 ÷ 3.

(2) Content of the previous lesson

In the previous lesson, students studied the following problem: "Suppose you want to divide 69 sheets of paper evenly between three people. How many sheets will each person get?" This made students think about how to solve 69 ÷ 3. They learned how to solve problems in which the numbers in both the tens and ones place of the dividend can be evenly divided by the divisor, 3. In problems like this, divide the numbers in each digit in the dividend by the divisor of 3, and then place those numbers in the quotient.

(3) Development of the current lesson

Understand the problem: The teacher writes the problem to be examined during the lesson on the chalkboard. The students think about the meaning of this problem while they copy it into their notebooks.

> Divide 72 sheets of paper equally between three people.
> How many sheets will each person get?

Since some children will not understand the problem from this text alone, the teacher confirms the question by showing the students seven packs of 10 sheets of colored paper (as shown at right) along with two individual sheets, and then explains, "the problem is to figure out how many sheets each person will get if you divide them up evenly between three people."

The teacher ensures that the children are aware of the following points:

- This problem is similar to the previous lesson as it is about "equal distribution among three people." That is, they can find the answer by using a "division" formula.
- This problem is different from the previous lesson which looked at 69 ÷ 3 in that 3 does not go evenly into the number in either the tens or the ones place not divide every into each of the digits which make up the number 72.

Thus, the teacher helps students recognize that the purpose of this lesson is to figure out how to solve 72 ÷ 3.

69 ÷ 3 = 23 (problem from the previous lesson)

Divide 69 sheets of paper equally between three people.
How many sheets will each person get?

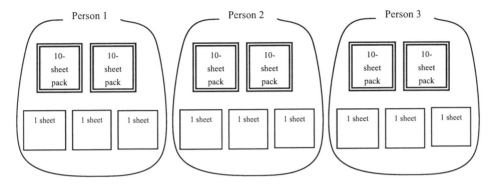

Current lesson

Divide 72 sheets of paper equally between three people.
How many sheets will each person get?

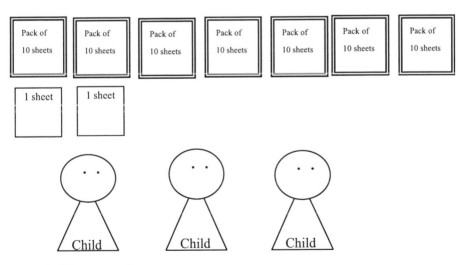

Goal of the current lesson

Think about how to compute 72 ÷ 3.

Developing a solution

Once the students understand the problem, it's time for them to devise their own plans for solving it. This is a time for students to work on their own. The teacher walks around the room, allowing the students to come up with ideas on their own, offering hints to students who are stuck, and encouraging students who have come up with a solution to think of other possible ways of solving the problem. The teacher then gives the students that come up with the typical methods of solving the problem a small chalkboard (or poster paper, white board, etc.) so that they can write out their solution large enough for presentation to the rest of the class to see.

Progression through discussion

The teacher places the ideas written out by the students on the small chalkboards on the main chalkboard, and has each student explain their approach.

The other students listen to the presentations while considering the following points, and then discuss their own ideas and questions.

- Can the method be used in any situation?
- Can it be done quickly?
- Is the explanation one that everyone can understand?
- How is it similar to other ideas? How is it different?
- What is unique about the idea?

For example, when students A to F have the ideas shown at right, the teacher has Students A, B, C, and D present their ideas.

The following type of exchange can be expected:

"A's approach is easy to understand because a figure is used in the explanation."

"If you turn A's approach into a formula, you get D's approach."

"B's approach is similar to the approaches of A and D."

"B, C, and D all split 72 into two parts."

A's Approach

Divide the sheets by unpacking one of the packs of colored paper.

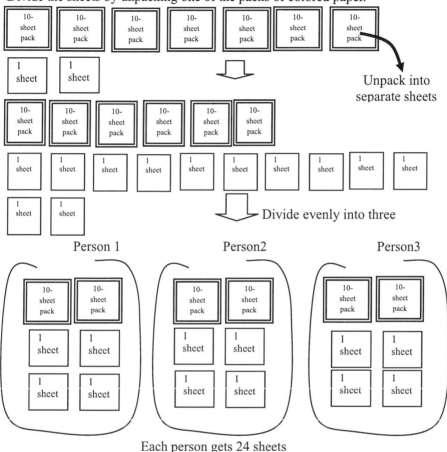

Each person gets 24 sheets

B's Approach

Compute by dividing 72 into 60 and 12, which are both divisible by 3.

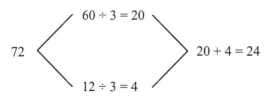

"Do the computation by dividing 72 into two parts, and then add the results at the end."

"C's suggestion of computing 70 ÷ 3 is a little hard. How did C compute that?"

"I thought about yesterday's lesson on 69 ÷ 3."

"It seems like B's approach could be used in any situation."

By reviewing the ideas developed by individual students with everyone, the class can come up with better ideas and learn to draw generalizations.

Summarizing

The teacher summarizes these ideas, reviews what has been learned during the lesson, and then reinforces the important points. The teacher should try as much as possible to summarize the lesson in the children's words.

In this example, the teacher might summarize as follows:

- You can split the number you want to divide into two numbers, perform the division, and then add the totals at the end. Or
- First, you divide the packs of 10 into three, and then take the sheets in the remaining pack, add it to the individual sheets, and divide those into three. Then combine the number of packs and individual sheets given to each person.

3. Problems with problem-solving oriented lessons

- Lessons are formal, and often do not follow the flow of the students' thinking processes.
- Students who quickly solve the problem during the individual study time may get bored. On the other hand, some students may be unable to understand the meaning of the problem itself or to come up with any possible method of solving it, thus spending the entire class period unable to do anything. This becomes a wasted time for both of these kinds of students.
- Students are not able to share the thinking process.
- If students have questions about their work, it is offen difficult to link those questions into a classroom discussion.
- It is not conducive to independent and autonomous activity by the students.

C's Approach: Perform the computation by dividing 72 into tens and ones.

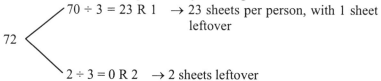

$70 \div 3 = 23$ R 1   → 23 sheets per person, with 1 sheet leftover

72

$2 \div 3 = 0$ R 2   → 2 sheets leftover

If you add the 1 and 2 leftover sheets, there are 3 leftover sheets. So then you can give one leftover sheet to each of the three people. This means each person gets $23 + 1 = 24$ sheets.

D's Approach: Divide the colored sheets into packs of 10 sheets and individual sheets.

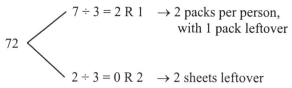

$7 \div 3 = 2$ R 1   → 2 packs per person, with 1 pack leftover

72

$2 \div 3 = 0$ R 2   → 2 sheets leftover

Add the remaining pack (10 sheets) plus the 2 individual sheets to get 12 sheets. Divide these equally among the three people, giving 4 sheets to each. Therefore, each person gets 2 packs plus 4 sheets, $20 + 4 = 24$ (sheets)

E's Approach: We learned in the last lesson that $69 \div 3 = 23$. Since three more sheets have added to this original number, one of these can be given to each person. So each person gets $23 + 1 = 24$ (sheets)

F's Approach: Divide 72 into two groups of 36, and then divide each of these by 3.

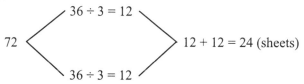

$36 \div 3 = 12$

72     $12 + 12 = 24$ (sheets)

$36 \div 3 = 12$

Section 2.2: Discussion-Oriented Teaching Methods and Examples

Discussion-Oriented Lessons for Improving Students' Expressive Skills

Hiroshi Tanaka

1. What is a "discussion-oriented" lesson?

Sometimes an idea that one cannot fully grasp on one's own becomes clear in the process of discussing it with a classmate or with the teacher.

In the process of carefully explaining one's ideas to a classmate, students may come to realize their own errors, thereby facilitating a greater understanding of the issue.

The act of discussing something is really the act of confirming one's own ideas, and is an effective means of improving students' academic skills.

For elementary school children, the process of making new discoveries and recognizing rules as they share their ideas with their friends is also crucial in terms of helping them learn important lessons about how to relate to other people as they move out into society.

Thus, discussion-oriented lessons not only are focused on cultivating mathematics skills, but also have important goals in terms of cultivating the students' humanity.

When children try to explain something that they understand, it becomes necessary for them to use various expressive skills that mathematics attempts to cultivate.

These include explaining things using figures, rephrasing ideas in simpler terms, and explaining things using examples.

It should be clear that the formal teaching of these methods does not nurture the ability to use them. It is precisely the student's desire to communicate something they know that cultivates the expressive methods that will be useful to them later.

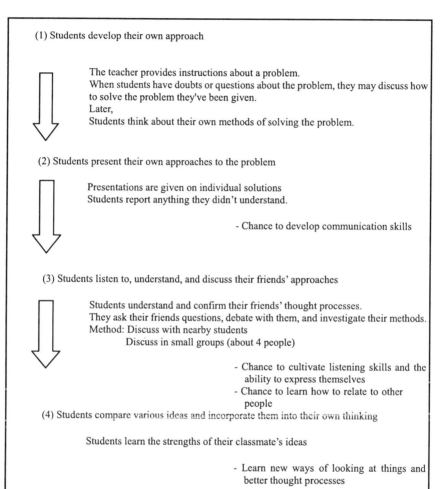

(1) Students develop their own approach

The teacher provides instructions about a problem.
When students have doubts or questions about the problem, they may discuss how to solve the problem they've been given.
Later,
Students think about their own methods of solving the problem.

(2) Students present their own approaches to the problem

Presentations are given on individual solutions
Students report anything they didn't understand.

- Chance to develop communication skills

(3) Students listen to, understand, and discuss their friends' approaches

Students understand and confirm their friends' thought processes.
They ask their friends questions, debate with them, and investigate their methods.
Method: Discuss with nearby students
        Discuss in small groups (about 4 people)

- Chance to cultivate listening skills and the ability to express themselves
- Chance to learn how to relate to other people

(4) Students compare various ideas and incorporate them into their own thinking

Students learn the strengths of their classmate's ideas

- Learn new ways of looking at things and better thought processes

Figure 1: Structural Model of a Discussion-Oriented Lesson

2. "Discussion-oriented" lessons in practice

2.1 Unit: *First Grade* "Comparison of Length"

The purpose here is to add to experiences that will form the foundation for students' understanding of quantities and measurement through activities like comparing the lengths of objects.

When comparing quantities like length, first allow students to experience the process of comparing length directly. Then create a situation in which direct comparison is not possible, and guide students to the concept that they can compare lengths by using the length of an object familiar to them based on how many lengths of the familiar object correspond to the length of the item being examined.

2.2 System of Curricular Content

Second Grade

Lessons on length are also taught in second grade. Here, based on their experiences in first grade, students deepen their understanding of length and learn to measure length in simple situations.

Specifically, students start measuring objects using universal units that were not used in first grade.

Here they learn about units of length: millimeters (mm), centimeters (cm), and meters (m).

Third Grade

In third grade, students learn not only about length, but also about the measurement of quantities such as dimensions and weight. Units of length up to the unit of the kilometer (km) are taught.

Also, when measuring these quantities, students are assigned activities in which they have to make "ballpark" guesses about the results and select the appropriate unit and measuring tool for a particular purpose.

These elements are presented on the opposite page, and are also summarized in Figure 2.

Direct Comparison: First Grade

When two objects whose lengths you want to measure can be moved, you can easily move the objects, line them up next to one another, and see which one is longer. In this case, students measure objects against a standard, learn to line both objects up carefully along one end, and compare the size of the objects along the other end.

Indirect Comparison: First and Second Grade

When the objects you want to compare cannot be moved, you cannot compare them directly. For example, this would be the case if you wanted to compare the height and width of a desk in the classroom. In this situation, you can compare them by copying the length of each side using pieces of string and using those to compare the lengths.

Comparison Using Arbitrary Units: First and Second Grade

When you want to know which of two objects is longer, select an appropriate item from nearby to use as a unit of measurement, and compare how many lengths of that item the object you want to measure is. Because this method can be converted into numbers for comparison, it is also a useful way to compare more than two objects simultaneously.

Comparison Using Universal Units: Second and Third Grade

In comparisons using an arbitrary unit, the size of the standard being used may differ depending on the person or pose other inconveniences, thereby making it necessary to use objective units of measurement. Teach students to perform measurements using those units.

Teaching Overview by Grade
First Grade: Direct comparisons, indirect comparisons, comparisons using arbitrary units
Second Grade: Shift from use of arbitrary units to universal units in comparisons
Units: cm, mm, m   Measuring instrument: Ruler
Third Grade: Measurement of longer lengths
Unit: km   Measuring instrument: Tape measure

Figure 2: Teaching System for the Study of Quantities Like Length

2.3 First Grade: Learning Length: Specific Elements of Lesson Development

Janken (Rock-Paper-Scissors) Race

Explain how the game works while affixing three strips of paper of differing lengths to the chalkboard.

In this game, if you win Janken with the Rock, you take the shortest strip of paper. If you win with the Scissors, you take the second strip. And if you win with the Paper, you take the third (longest) strip. You can proceed as long as the strip of paper you win.

Rock   ▢

Scissors ▢

Paper   ▢

These strips of paper are actually 10, 20, and 30 cm long, but for this game, the students should not know the relationship between the lengths of the strips. Thus, their edges should not be lined up when the game is being explained. One of the aims of this game is for the students to become aware of the length of each strip of paper in the process of playing the Janken Race.

"The teacher's course is shorter!"

Start the game.

First, the teacher should play the game with a student representing the whole class. This has the benefit of ensuring that all the students know the rules of the game and of allowing everyone to think about and figure out any problems that might arise during the game, and thus of developing learning.

Because the students are first graders, it is essential that they learn through playing games and having fun. However, this is a way to enable students to discover common problems during the game and to solve them by talking about the problems with one another and the main aim is not for the students to play.

Show the lanes to be used for the game to the children. First, show the students' lane. Then draw the teacher's lane a short distance away from it on the chalkboard.

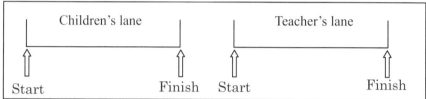

"The teacher's course is too short!"

"You should copy it using your hand this way!"

"What if you were to check it using your finger like an inchworm?"

"That's not very precise, is it?"

The children make a big racket as soon as they are shown the lanes. This is because the teacher's course is clearly shorter than the students'. For this game, we made the children's course 1m and the teacher's course 80 cm long.

> Child A: "Teacher, that's cheating. No matter how you look at it, your course is shorter."
>
> Teacher: "I don't think so. Maybe you're just seeing it wrong."
>
> Child B: "If you check it by using your arms, you'll be able to tell right away."
>
> Teacher: "Okay, try it."

Child B comes up, holds his hands out to touch both ends of the course on the chalkboard, and then moves over to the adjacent course to compare them. Watching this, Child C says, "No way. The distance between your hands is getting shorter. You have to do it right." Child B tries several times to fix it, but the results are still imprecise. Child D then says, "Teacher, what if you were to check it instead by using your finger like an inchworm?" They tried this too, but they realized there was a problem when the distance between their fingers changed during the process.

In this way, the students exchanged ideas with one another, determined that they need something whose width would not change, and came up with the idea of using the strips of paper used for the Janken Race. Thus, the lanes were adjusted to the same length.

"Who is the Janken Race Champion?"

Many of the children who watched the end of the Janken Race between the student and teacher said they wanted to do it, too. So this time, they decided to do a Janken Race with everyone participating. They created teams of two and had them spread out in the classroom for the challenge. This time, the rules of the game were changed slightly, on the premise that it would be more fun to see who could come up with the longest length of paper than to create equal lanes. Fun games were played all around the room. When the game ended, the students looked at their lined up strips, and determined which member of each competing pair had won. Because this was done using direct comparison, the students understood quickly.

"How many lengths of this strip of paper are there? You should check."
"That makes the courses equal." "The Janken Race begins!"

"We want to do the Janken Race, too!" "Let's play in groups of two."

"I did it! I won!" "But who is the class champion?"

When the teacher then asked, "Who is the class champion?" the students' faces fell. Child A said, "If we were going to do it that way, we should have lined them all up at the start line at the beginning." Child B said, "It will be pretty tough to line them all up now," and Child C chimed in, "If you tell us how many strips there are, couldn't we compare them using the strips of paper? For example, what if we check using the longest Paper strips?" Then Child D continued with the idea, "If we do this, we should use the shortest strip because we might end up with a half strip." Upon comparing the Paper strip to the Rock strip this way, the students learned that the strips were in proportional length, 1:2:3. "Teacher, this strip, the Paper strip, is three times the size of the Rock strip." "In that case, if we say the Rock strip is 1 point, the Scissors strip is 2 points, and the Paper strip is 3 points, we can add them up to compare the lengths achieved by each student."

Thus, the first grade students were able to help one another correct the weaknesses in their own approaches by discussing their ideas with their classmates, and were able to solve the problem. The children actually expressed a lot of ideas that mathematics tries to teach as they spoke naturally about what was going on during this activity.

The role of the teacher is to organize these concepts and make students aware of them. I have actually conducted research on lessons that focus on the children's "first responses" that occurred during this activity (Hiroshi Tanaka (2001)). Two of the typical words that came up were "for example," and "if," words that lead to generalizations. After the phrase "for example," the child would express their own specific "way of understanding" the issue. When a child wants to convey something they understand to their friends in an easy-to-understand way, using specific scenarios. The word "if" is used when one wants to make a generalization by thinking about a development, changing the conditions, or providing a counterexample. These are also words that cultivate skills of inductive inference, a typical form of reasoning.

Reference
Hiroshi Tanaka (2001), *Lessons that Cultivate Skills of Mathematical Expression* [Sansuuteki hyougenryoku wo sodateru jugyou], Toyokan.

"Which strips should we use to check how many lengths is?"
"Teacher, I discovered something interesting."
"The Rock strip is the same length as two of the Scissors strips."
"The Paper strip is three times as long as the Rock strip."

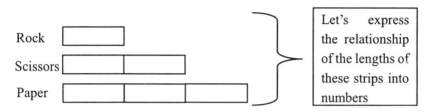

Ex.) If the lengths of Child A's strips are assigned to scores:

Rock	Paper	Scissors	Rock

If you think about how many lengths each of these represent:

1 point	3 points	2 points	1 point

You can add them up to get 7 points: $1 + 3 + 2 + 1 = 7$

Two phrases that convey inductive inference, which we want elementary school students to learn:
"For example": The ability to speak using many examples
"If": The ability to generalize by giving counterexamples or changing the conditions.

Children's expressive skills and mathematical thought processes
   Relationship between the children's "first responses" and mathematical thought processes.

Section 2.3: Problem-Discovery Oriented Teaching Methods and Examples
What are Problem-Discovery Oriented Lessons?

Yoshikazu Yamamoto

1. What is a Problem-Discovery Oriented Lesson?

Unlike lessons in which students solve learning problems that the teacher presents to them, a problem-discovery oriented lesson is one in which students become aware of a learning problem that they have to solve themselves as they proceed through a learning activity, and then work on solving that problem. In other words, the structural model of the lesson consists of three stages, as shown in Figure 1: (1) initial learning activity, (2) discovery of a problem that must be solved, and (3) solution of the problem.

Thus, the initial activity the students undertake has to be the kind of activity that raises their awareness of a problem. The teacher determines this initial activity by trying to predict the reactions of the students at the stage of the lesson design. Thus, the teacher must try to orchestrate steps in the activities so that the students can find a way to solve the problems that they will discover during the activity. Problem-discovery oriented lessons emphasize the children's change of awareness, thereby it is a type of lesson that challenges the teacher's teaching skills.

In these kinds of problem-discovery oriented lessons, the children's learning is not a matter of "being taught" by the teacher, but a matter of "learning" on their own. Thus, these lessons are not only significant in enabling students to learn mathematics, but are also significant in effectively making them aware of the significance of the learning process itself.

(1) Initial learning activity

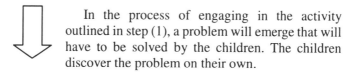

The teacher decides on an activity orchestrated ahead of time to elicit the learning problem for the lesson, but it is not one that students will be able to directly recognize as a learning topic. How these activities are created is the key to the success of problem-discovery oriented lessons.

(2) Discovery of a problem that must be solved

In the process of engaging in the activity outlined in step (1), a problem will emerge that will have to be solved by the children. The children discover the problem on their own.

(3) Solution of the problem

The problem discovered in step (2) is a problem that the students feel the need to solve. The students use their existing knowledge and thought processes to solve the problem.

Figure 1: Structural Model of a Problem-Discovery Oriented Lesson

In a problem-discovery oriented lesson, the teacher has to orchestrate the lesson so that the predetermined problem emerges during the initial activity. This is the component in which the teacher's skills are especially important.

The learning approach in a problem-discovery oriented lesson sees the child as the primary agent in learning, not as one who learns a lesson content taught by the teacher.

2. Problem-Discovery Oriented Lessons in Practice

2.1 Unit: 6th Grade "Addition and Subtraction of Fractions"

2.2 Connecting Previous Lessons to the Current Lesson
[Fourth Grade]

Children first encounter fractions in fourth grade, when they learn what fractions are, and also the meanings of terms like "fraction," "denominator," and "numerator." In Japanese, it is important to know that a fraction, such as one third, is read and therefore understood as "*san bun no ichi*" which is the reverse of the English order of expression. The "*san bun*" refers to the denominator, and the "*ichi*" to the numerator. The expression "*san bun no ichi*" refers to something equivalent to "dividing by three and then take one part". When Japanese children write fractions, it is normal for them to write the denominator first, then to write the vinculum and the numerator last of all. This Japanese form of naming fractions does not allow the possibility of talking incorrectly, for example, about one third or two fifths as "one over three" or "two over five".

Children also learn that a fraction in which the numerator is less than the denominator is called a "proper fraction," while a fraction in which the denominator is less than or equal to the numerator is called an "improper fraction." These are fraction expressions that have the advantage of making it easy to understand how many unit fractions are put together.

They also learn how to express fractions larger than 1, as "mixed fractions," such as $1^2/_5$. Because the fraction is expressed as a combination of both a whole number and a proper fraction, this format has the advantage of making it easy to figure out the size of the fraction.

The advantages of both improper fractions and mixed fractions are learned and completed in fourth grade.

[Fifth Grade]

In fifth grade, students will learn how to add and subtract proper fractions based on what they have learned about fractions in fourth grade. The addition and subtraction they learn up to the fourth grade is limited in its application to whole numbers and decimals, but this will be expanded to fractions in fifth grade. That is, the purpose of the fractions lessons taught in fifth grade is to expand the range of numbers to which students can apply addition and subtraction. Lessons also address the process of performing addition and subtraction limited to the fractions with a common denominator.

System of Teaching Fractions in Japan

Fourth Grade Curriculum

Regarding fractions in partition

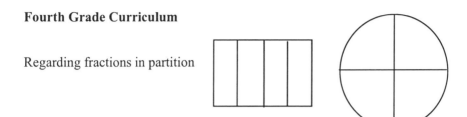

Fractions that express fractional quantities

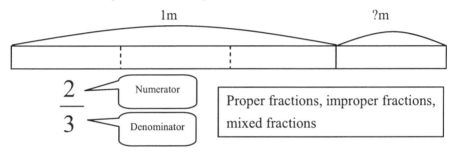

$\dfrac{2}{3}$ Numerator — Denominator

Proper fractions, improper fractions, mixed fractions

Fifth Grade Curriculum

Addition and subtraction of fractions with a common denominator

$$\frac{2}{5} + \frac{1}{5} = \frac{3}{5} \qquad\qquad \frac{5}{7} - \frac{2}{7} = \frac{3}{7}$$

Fractions of equal size

$$\frac{1}{2} = \frac{2}{4} = \frac{3}{6} = \frac{4}{8} \cdots\cdots$$

Fraction division

$$a \div b = \frac{a}{b}$$

Lessons also address fractions of the same size, like 1/2, 2/4, and 3/6, but terms such as "reduction to a common denominator" and "reduction to lowest terms" and the addition and subtraction of fractions with different denominators are part of the sixth grade curriculum.

In fifth grade, the curriculum includes a new perspective toward fractions, so-called "quotient fractions." Thus far, the division of [a ÷ b] had only been taught using the range of whole numbers or decimals, but students will now learn that a quotient that cannot be divided evenly can be expressed as a fraction [a/b], based on the concept of quotient fractions.

Sixth Grade

The sixth grade curriculum in fractions includes the addition and subtraction of fractions with different denominators and the multiplication and division of fractions.

Here we will examine a lesson on the addition and subtraction of fractions with different denominators as an example of a problem-discovery oriented lesson. The students learned how to add and subtract fractions with common denominators in fifth grade. Developing a method of adding fractions with different denominators on their own will therefore be meaningful to them. The teacher is not to teach this idea. In the process of developing this idea, the students will discover the following two problems:
- Is it possible to add and subtract fractions with different denominators?
- How do I add and subtract fractions with different denominators?
This example demonstrates a lesson in which the children will discover these kinds of problems and solve them.

2.3 Purpose: To cultivate perspectives on adding and subtracting fractions with different denominators through an activity that involves disassembling and assembling pattern blocks.

Sixth Grade Curriculum

Addition and subtraction (reduction to common denominator, reduction to lowest terms) of fractions with different denominators

$$\frac{2}{5} + \frac{1}{2} = \frac{4}{10} + \frac{5}{10} = \frac{9}{10}$$

$$\frac{5}{7} - \frac{1}{3} = \frac{15}{21} - \frac{7}{21} = \frac{8}{21}$$

Multiplication and division of fractions

$$\frac{3}{4} \times \frac{1}{2} = \frac{3 \times 1}{4 \times 2} = \frac{3}{8}$$

$$\frac{4}{9} \div \frac{3}{5} = \frac{4 \times 5}{9 \times 3} = \frac{20}{27}$$

Process of the Children Eliciting Problems by Themselves in this Example

○ Introduce the addition and subtraction of fractions with different denominators (*Janken* Game)

○ Is it possible to perform calculations using fractions with different denominators?

○ How do I add or subtract fractions with different denominators?

2.4 How Children Learn in this Lesson

1. Introduction

The teacher says, "Let's play the *Janken* (rock scissors paper) Game using these pattern blocks," and prepares the pattern blocks shown in Figure 2 in front of each child.

If pattern block (a) represents one *whole block* (*whole block* is a randomly selected unit, anything can be used here), then the total pattern blocks prepared is 4 *whole blocks* per person. Confirm that pattern blocks (b), (c), and (d) each represent 1/2 of a *whole block*, 1/3 of a *whole block*, and 1/4 of a *whole block*.

The rules of the game are shown at right.

At this point, some students will ask, "So doesn't this mean that we don't need the (a) block?" The teacher responds by saying, "That's a good question. But first let's just try it and see," and then starts the game. The classroom will then be filled with happy faces of the children having fun playing *Janken*.

2. Problem Discovery 1 through an Activity (Subtracting fractions with different denominators)

After a little while, some children will run out of their (b) and (c) pattern blocks. At this point, the teacher temporarily stops the game and make the children think of how to play the game if they run out of all their (b) and (c) blocks. This explanation makes the children realize how they can use block (a) by "taking block (a) and getting change for it." This also explains to the children why they needed to have block (a). Express the idea and understand the mathematical concept of "giving change" using formulas: [1 - 1/2], [1 - 1/3], and [1 - 1/6].

Some students will say, "In that case, when I have to give a "(c)" to my opponent, I can also give him a "(b)" and ask for change." As they think about this idea, and express it as a formula, they get [1/2 − 1/3], and realize that they have to subtract fractions with different denominators. At this point, some of the children will realize the problem: "Can we subtract fractions with different denominators?"

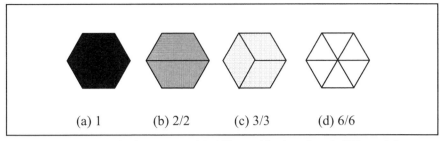

(a) 1   (b) 2/2   (c) 3/3   (d) 6/6

Figure 2: Pattern Blocks Prepared (Different block colored differently)

(1) Play *Janken* (Rock Scissors Paper) with several classmates, one after another.
(2) If you win with Rock, take a block (b) (1/2 *of a whole block*) from your opponent. If you win with Scissors, take a block (c) (1/3 *of a whole block*) from your opponent, and if you win with Paper, take a block (d) (1/4 *of a whole block*) from your opponent.

Figure 3: Rules of the Initial Activity (*Janken* Game)

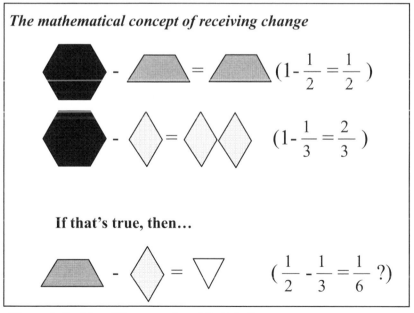

Figure 4: Problem Discovery 1 and its Solution
(Subtracting Fractions with Different Denominators)

This is the moment when a problem emerges among the children.

However, some children will say, "I can calculate it. The answer is 1/6." While some students who intuitively say the answer is 1/6, others will express the idea that will lay the groundwork for the concept of reducing fractions to a common denominator, saying, "Shouldn't we exchange one (b) *(1/2)* for three (d)'s *(3/6)*, and then exchange one (c) *(1/3)* for two (d)'s *(2/6)*?" They will visually start to see these things as they manipulate the pattern blocks. This is the result of the characteristics inherent in these blocks to have worked effectively and is the reason that these pattern blocks are used in this exercise.

After checking these ideas, restart the game.

3. Problem Discovery 2 through an Activity (Adding fractions with different denominators)

At the appropriate time, signal the students to stop the game. Without the teacher saying anything, the children will start to try to figure out how many *whole blocks* their pattern blocks add up to.

As they do this, they will find that pattern blocks (b), (c), and (d) are all different types, and don't add up to "a *whole block*." Ultimately, the students will come up with the idea of trying to create a *whole block* by combining different types of pattern blocks, as shown at right. They will also find ways that different types of pattern blocks, like (b) *(1/2)* and (c) *(1/3)*, can be mixed within other fractions. In other words, they will add fractions with different denominators.

When the teacher addresses this and expresses it as a formula, the students will realize that this is a method of adding fractions with different denominators. And based on what they learned earlier about subtracting fractions with different denominators, they will confirm that the answer is 5/6.

This also utilizes the mathematical concept of exchanging money.

To make the exchange
Replace them with the (d) pattern blocks

$$\triangle\triangledown\triangle \ - \ \diamondsmall \ = \ \triangledown \quad (\frac{3}{6}-\frac{2}{6}=\frac{1}{6})$$

Problem Discovery 2 and Its Solution (Adding Fractions with Different Denominators)

Approaches to making "1 *whole block*" (method)

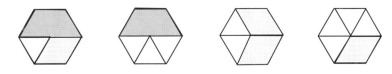

These are all examples of adding fractions with different denominators.

Dealing with pieces other than "1 *whole block*"

$$\text{(trapezoid)} + \diamondsmall = \text{(shape)} \quad (\frac{1}{2}+\frac{1}{3}=?)$$

⇩

To make the exchange
Replace all three shapes above with the (d) pattern blocks

$$\triangle\triangledown\triangle + \diamondsmall = \triangle\triangledown\triangle\triangledown \quad (\frac{3}{6}+\frac{2}{6}=\frac{5}{6})$$

CHAPTER 3

Trends of Research Topics in Japan Society of Mathematical Education

Lesson Study in Elementary Schools

Section 1.1: What are the Features of Lesson Study Projects Conducted in Elementary School Mathematics Departments?

Tadayuki Kishimoto & Kozo Tsubota

Lesson Study Policies

Mathematics Lesson Study in Japan is characterized by the principles described in Figure 1. The educational policies are to encourage students to take an active role in constructing mathematics problems by communicating with one another, and to develop their belief in their ability to learn and think on their own. These policies appear to have much in common with what is known as social constructivism in the USA and Europe, and has long been utilized in Japan. The principle in the area of academic competence in Japan is to focus on expanding students' mathematical ways of thinking and their mathematical interest, eagerness, and attitudes, as well as their knowledge or skills. The policies in the realm of teaching contents emphasize the importance of the basics and fundamentals, as well as integration and development of knowledge. These are well suited to the educational policies above. These policies also include building on previous lessons, connecting questions, and developing open-ended problems. The policies in the area of evaluations are to encourage the use of standards-based assessment rather than ranking students, and to shift the focus away from what students "cannot do" to what they "can do."

Problem-Solving Oriented Lessons

Many Japanese lesson plans can be described as "problem- solving oriented lessons." These consist of four separate stages in which students (I) understand the problem, (II) make a plan, (III) execute the plan, and (IV) evaluate their solutions. Understanding the problem involves building on previous lessons. In the plan-making stage, the teacher helps students to develop possible solutions to the problem. In the execution stage, the focus is on independent problem-solving. And in the evaluation stage, the teacher summarizes and discusses diverse ideas and strives to tie the current lesson to the next one. Different learning formats, such as individual learning, group learning, and full-class learning, are used in different stages, as appropriate. Small group instruction and team teaching may also be utilized in certain situations.

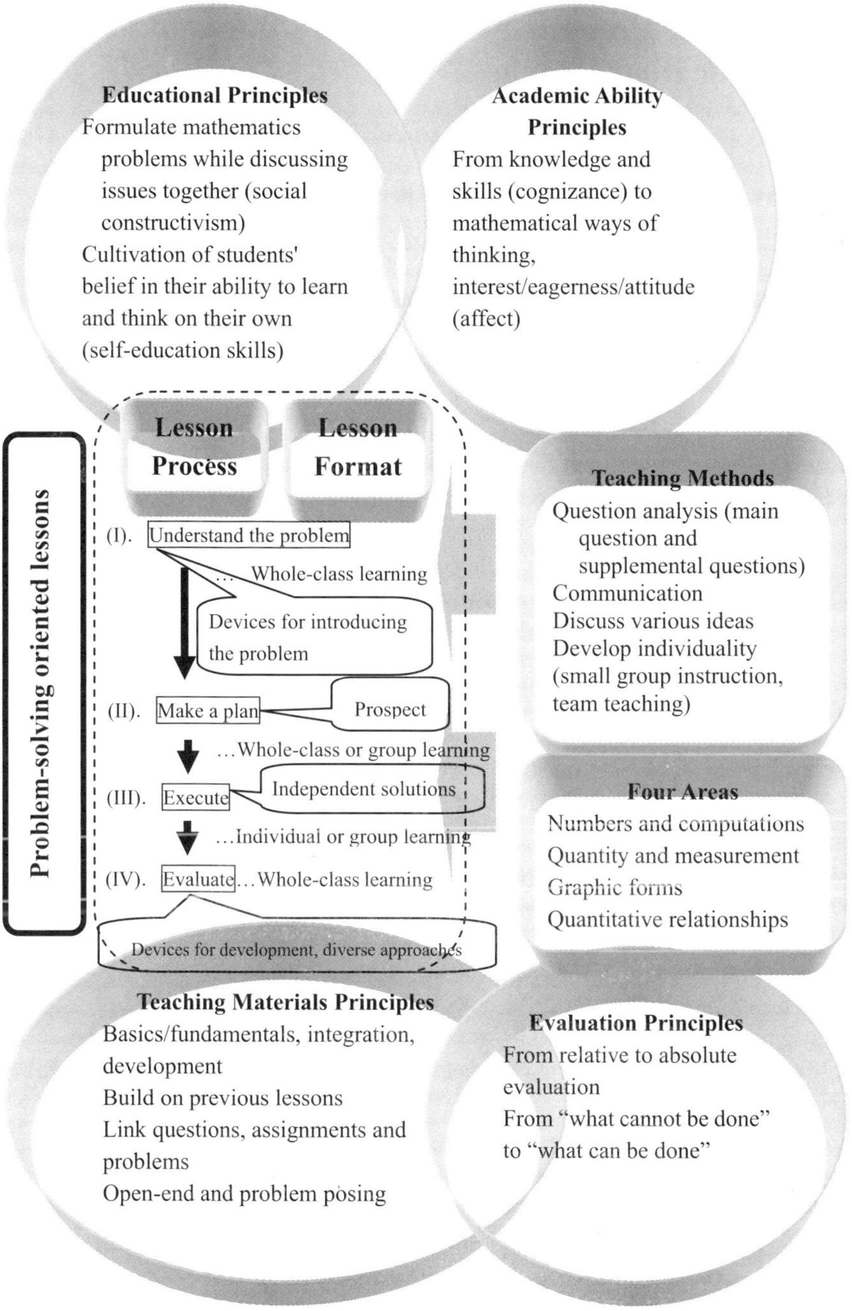

Educational Principles
Formulate mathematics
problems while discussing
issues together (social
constructivism)
Cultivation of students'
belief in their ability to learn
and think on their own
(self-education skills)

**Academic Ability
Principles**
From knowledge and
skills (cognizance) to
mathematical ways of
thinking,
interest/eagerness/attitude
(affect)

**Lesson
Process**

**Lesson
Format**

Problem-solving oriented lessons

(I). Understand the problem
...Whole-class learning
Devices for introducing
the problem

(II). Make a plan — Prospect

...Whole-class or group learning

(III). Execute — Independent solutions
...Individual or group learning

(IV). Evaluate ...Whole-class learning

Devices for development, diverse approaches

Teaching Methods
Question analysis (main
question and
supplemental questions)
Communication
Discuss various ideas
Develop individuality
(small group instruction,
team teaching)

Four Areas
Numbers and computations
Quantity and measurement
Graphic forms
Quantitative relationships

Teaching Materials Principles
Basics/fundamentals, integration,
development
Build on previous lessons
Link questions, assignments and
problems
Open-end and problem posing

Evaluation Principles
From relative to absolute
evaluation
From "what cannot be done"
to "what can be done"

Figure 1: Special characteristics of mathematics Lesson Study in Japan

Section 1.2: How Have the Goals of Mathematics Education Changed?

Tadayuki Kishimoto & Kozo Tsubota

Post-War Revisions to the National Course of Study

According to Article 18, Clause 5 of the School Education Law, the goals of mathematics education are "To ensure an accurate understanding of the mathematical relationships necessary in everyday life and to cultivate the ability to perform mathematical processes." These were reflected in the "National Course of Study for Mathematics (Draft)" (1947), the "Mathematics Contents Table" (1948), and the "National Course of Study in Mathematics for Elementary School" (1951). These guidelines focused on practical life lessons, and therefore emphasized the cultivation of the ability and willingness to use mathematics in everyday living. The "National Course of Study in Mathematics for Elementary School" was revised in 1958, at which time their legal authority was established. These guidelines focused on systematic learning, and aimed toward the systematization of educational content while emphasizing "mathematical thinking." Revisions made in 1968 were especially focused on modernization, and emphasized the concepts of sets and functions even at the elementary school level. In 1977, the curriculum revisions aimed to 'comfort and enrichment', and emphasized the careful selection of educational content as well as the educational basics/fundamentals. The guidelines revised in 1989 featured new perspectives on academic ability, focused on the benefits of mathematics, and emphasized raising the affective elements of mathematics education. Revisions made in 1998 highlighted the development of 'students' zest for living'. The curricular content for the new five-day school week was also carefully selected, and mathematical activities and basics/fundamentals were emphasized.

Curricular Shifts Between Grade Levels in the National Course of Study

Shimizu, S. (2000) summarized the shifts in curricular content between grade levels enacted in the revised National Course of Study (Table 1, partial revision). After World War II, the curriculum delayed more than one year. After returning to the pre-war schedule on the 1958 revisions, others contained relatively few changes. However, the 1998 revisions shifted the start of the school year delay again due to the careful selection of curricular content for the new five-day school week system.

Table 1: Shifts in the Curriculum Between School Years and Schools (Shimizu, S. 2000)

Subject	1951	1958	1972	1977	1989	1998
Introduction to fractions	2	→	→	3	→	4
Introduction to decimals	3	→	→	→	→	4
Addition	2	1	→	→	→	→
Subtraction	2	1	→	→	→	→
Multiplication	3	2	→	→	→	→
Adding fractions with unequal denominators	7	5	→	→	→	6
Multiplication of fractions	7	6	→	→	→	→
Squares, rectangles	4	3	2	→	→	3
Types of triangles	7	5	3	→	→	4
Types of quadrangles	7	5	4	→	→	5
Congruent figures	8	→	4	5	→	8
Line symmetry, point symmetry	9	7	5	6	→	7
Enlarged and reduced figures	5	6	→	→	→	9
Cubes, cuboids	5	4	→	→	→	6
Prisms, cylinders	5	6	→	→	→	→
Area of squares and rectangles	5	4	→	→	→	→
Area of triangles and quadrangles	8	5	→	→	→	→
Area of circles	8	5	→	→	→	→
Volume of cubes and cuboids	6	4	→	5	→	6
Proportions	8	6	→	→	→	→
Reverse proportions	8	6	→	→	→	7
Introduction to symbols/algebra	7	→	5	→	→	7

The years indicate the years in which the National Course of Study was revised.

Section 1.3: How Have Research Trends at the Japan Society of Mathematical Education National Conference Changed?

Tadayuki Kishimoto & Kozo Tsubota

Lesson Study Trends in Japan

Table 1 shows the number of presentations given at the National Conference of the Japan Society of Mathematical Education (Kindergarten and Elementary School Division). Every year the conference features about 150 research presentations, indicating a high level of Lesson Study activity. The overall trends in the presentation topics indicate that in the area of teaching goals, few studies are focusing on knowledge and skill proficiency, while numerous studies are being conducted on promoting mathematical ways of thinking and mathematical interest, eagerness, and attitudes. There is also a relative abundance of studies on lessons that encourage student initiative and develop students' individuality.

Recent Trends in Lesson Study

Tables 2 and 3 show trends in the keywords used in presentation topics over the past 10 years. The category "initiative" refers to the use of phrases like "learn on their own" or "think on their own." The category "develop individuality" refers to the use of phrases like "suited to the individual" and "individual differences." "Small group instruction" includes references to "instruction by level of proficiency."

The general trend over the past ten years indicates that keywords like "basics/fundamentals," "fun," and "small group instruction" are being used with increasing frequency in research presentations, while keywords like "benefits," "eagerness," "mathematical ways of thinking," and "individual students" are becoming less common. It has been suggested that these kinds of changes are a result of the 1998 revisions to the National Course of Study (NCS). That is, the 1989 NCS emphasized the benefits of mathematics, thereby triggering numerous many research projects on "the benefits of mathematics." The 1998 NCS, by contrast, emphasizes the principle of developing individuality based on equity, thereby triggering an abundance of research on "basics/fundamentals," "fun," "small group instruction," and "developmental and supplemental leaning."

Table 1: Trends in the Number of Research Presentations at the Japan Society of Mathematical Education National Conference (Kindergarten and Elementary School Division)

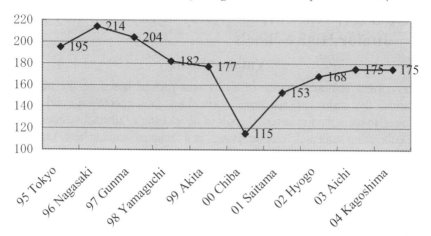

Table 2: Trends in the Research Themes of the Kindergarten and Elementary School Division I (Teaching Goals)

	1995	1996	1997	1998	1999	2000	2001	2002	2003	2004
Basics/fundamentals	1	2	1	0	1	1	3	12	10	22
Eagerness	26	28	18	6	4	8	1	12	7	0
Initiative	15	35	17	23	21	18	16	21	19	19
Enjoyment	1	2	4	2	8	9	18	20	29	16
Benefits	18	16	18	11	4	5	7	5	25	7
Mathematical ways of thinking	9	13	18	15	11	4	4	2	7	7

Table 3: Trends in the Research Themes of the Kindergarten and Elementary School Division I (Teaching Methods)

	1995	1996	1997	1998	1999	2000	2001	2002	2003	2004
Developmental learning	0	1	0	1	0	0	0	2	0	11
Develop individuality	5	6	6	2	5	3	4	3	4	5
Small group instruction	0	0	0	0	0	0	0	4	9	11
Take caring each	11	16	13	18	8	4	5	4	2	7

Lesson Study in Junior High Schools

Section 2.1: The Current Status of Lesson Study in Junior High Schools

Yutaka Oneda

The current National Course of Study was revised based on the needs of students, the status of academic curriculum implementation, and social changes, to meet the fundamental goals of developing a "distinctive education" and cultivating students' "zest for living" in a "relaxed" atmosphere. Later, the Ministry of Education, Culture, Sports, Science and Technology further revised those guidelines, enacting policies for "promoting learning," "pioneering improvements in academic skills," and "creating teaching materials that focus on developmental and supplemental learning," clarifying the "minimum standards" in the National Course of Study, and eliminating regulations in the NCS which stops teachers to teach more than standards.

In response to these trends, junior high school Lesson Study has been realigned toward cultivating students' ability to think on their own, ensuring the acquisition of basic knowledge while developing more relaxed educational activities, and aiming to enhance education that develops students' individuality. These realignments are addressed in the following four studies:

Study on the Facets of Student Understanding for Each Learning Topic

Given that students have varying skills, aptitudes, and interests, teachers must know their students individually in order to ensure that they can all achieve the same educational goals. This is a basic practical research project for learning the facets of student understanding.

Study on Teaching Methods and Teaching System Devices

Practical research on teaching systems, like lesson formats that include whole-class teaching, individual teaching, and group teaching. Practical research on effective teaching methods that meet students' needs and are suited to the teaching environment, such as repetitive teaching, teaching by level of proficiency, and subject-based teaching.

Development of Teaching Materials that Promote Mathematical Activities

Practical case-study research using teaching materials that promote mathematical activities like thinking on one's own and striving to learn by pursuing one's interests.

Personalized Teaching and Evaluation Methods

Practical research on appropriately evaluating student learning, using that information for improving teaching, reevaluating, and promoting the integration of teaching and evaluation.

Table 1: Overview of results from the 2001 Elementary and Junior High School Academic Curriculum Implementation Survey (Teacher Questionnaire)

Are you implementing team teaching or small-group teaching? (8th grade)

Category	Yes, in most classes	Yes, to some extent	Not to any significant degree	No, virtually not at all	No response
Mathematics	17.1%	9.2%	16.8%	56.2%	0.0%

Are you teaching lessons formulated for learning groups, based on their level of proficiency?

Category	Yes, in most classes.	Yes, to some extent	Not to any significant degree	No, virtually not at all	No response
Mathematics	4.1%	3.2%	17.6%	74.4%	0.7%

Are you teaching lessons that incorporate developmental topics?

Category	Yes	Yes, to some extent	Not to any significant degree	No, not at all	No response
Mathematics	10.8%	38.9%	37.4%	12.0%	0.8%

Do you provide additional instruction to students who are having problems understanding the material during breaks or after class?

Category	Yes	Yes, to some extent	Not to any significant degree	No, not at all	No response
Mathematics	10.5%	40.4%	37.8%	10.4%	0.9%

Section 2.2: Changes in Curriculum and Class Hours in the New National Course of Study

Yutaka Oneda

Changes in Curriculum and Class Hours

Phrases like "shift away from mathematics and science," "decline in academic skills," and "the harmful effects of relaxed education" are appearing in the newspapers. This can be attributed to the reductions in the curricular content and the decline in the number of class hours that have been enacted through revisions to the National Course of Study since the 1980s. The guidelines enacted in 2002 in particular introduced "the five-day school week," "significant reductions to the curriculum," and the introduction of "the integrated-study period." They also significantly reduced the curricular content and class hours for elementary and junior high schools. The current nine years of compulsory elementary and junior high school education have approximately two fewer years worth of lessons than the curriculum that was implemented in the 1970s.

However, these revisions were not introduced with the goal of implementing new concepts in mathematics education. Nevertheless, the shift from "relaxed educational activities" to "fewer class hours" was a result of national policies, and unfortunately, the topics were reconfigured as a result. Much of the content shifted from the lower to the upper levels, and high schools are now concerned that they have too much material to get through.

The following table shows the changes in the class hours and the changes in the content of the curriculum.

Table 1: Changes in the number of mathematics class hours (annually)

School	Elementary school				Junior high school				Senior high school
Year	Lower years	Middle years	Upper years	**Elementary total:**	7th grade	8th grade	9th grade	Junior high total:	Mathematics I
1970	242	385	420	1047	140	140	140	420	6 units
1980	311	350	350	1011	105	140	140	385	4 units
1990	311	350	350	1011	105	140	140	385	4 units
2000	269	300	300	869	105	105	105	315	3 units

Changes in the curriculum (material eliminated or shifted and integrated) (2000)

Eliminated content
Parallel, rotational, and line-symmetric displacement (7th grade)
Cross-sections of three-dimensional figures, projection (7th grade)
Figures that fulfill conditions (7th grade)
Mathematical expressions (approximate values, the binary numeration system, flow charts) (8th grade)
Square root tables (9th grade)

Shifted and integrated content
Shifted to higher grades
Similarity of figures (8th grade) to 9th grade
Shifted to lower grades
Circles and straight lines (tangent lines) (9th grade) to 7th grade
Arc length and area of a sector (9th grade) to 7th grade
Some probability lessons (9th grade) to 8th grade

Shifted to upper high school
Number sets and the four operations of mathematics (7th grade) to Mathematics I
Linear inequalities (8th grade) to Mathematics I
Terms for rational and irrational numbers (9th grade) to Mathematics I
Formulas for solving quadratic equations (9th grade) to Mathematics I
Centroid of a triangle (8th grade) to Mathematics A
Some properties of circles (e.g., the properties of two circles) (9th grade) to Mathematics A
Area and volume ratios of similar figures (9th grade) to Mathematics I
Spherical volume, surface area (9th grade) to Mathematics I
Data and sampling surveys (9th grade) to Mathematics Basics, Mathematics B, Mathematics C
Various events and functions (9th grade) to Mathematics I

Section 2.3: Research Trends at the Japan Society of Mathematical Education National Conference

Yutaka Oneda

Research Trends at the National Conference

Research trends at the national conference tended toward expanding the abilities and potential of each student, striving to enhance education for developing individuality, enhancing student-initiated learning, and striving to improve teaching techniques based on the needs of individual students. Academic and practical research on teaching methods that stimulate students' interests and promote active, inquisitive learning while developing each student's individuality through problem-solving oriented lessons is now in the mainstream.

There is a need for studies on how to evaluate "mathematical approaches and ways of thinking" and "interest / eagerness / attitude," and for practical research that aims to strike a balance between teaching practices that focus on mathematical activities and teaching practices designed to ensure the acquisition of basic knowledge.

Sessions at the National Conference

The 2003 national (Aichi) conference featured the following 17 sessions, as well as a poster session. New sessions on "basic academic skills" and "small-group teaching" were offered in response to changing social trends. Over the past several years, there has been an increase in the number of presentations on teaching methods, evaluations, basic academic skills, and small-group teaching.

> Academic Curriculum, Educating Handicapped Students, Numbers and Formulas, Figures, Quantity Relationships, Problem-Solving and Subject-Based Learning, Mathematical Ways of Thinking, Teaching Methods, Computers and Educational Equipment, Evaluations, Integrated Learning, Basic Academic Skills, Small-Group Teaching, Basic and Student-Guided Study, poster session.

Table 1: Number of presentations in each session at the National Conference

Session	2003	2002	2001	2000	1999	1998
Academic Curriculum	5	10	9	7	8	7
Educating Handicapped Students	3	0	1	0	0	0
Numbers and Formulas	8	6	5	5	8	8
Figures	23	14	22	17	20	16
Quantity Relationships	8	12	5	7	8	12
Problem-Solving and Subject-Based Learning	17	15	17	18	13	22
Mathematical ways of thinking	14	14	21	12	14	23
Teaching Methods	16	19	32	23	24	32
Computers and Educational Equipment	4	8	11	9	16	7
Evaluations	12	7	6	4	8	7
Integrated Learning	1	4	6	5	14	-
Basic Academic Skills	8	-	-	-	-	-
Team Teaching and Small-Group Instruction	20	-	-	-	11	6
Basic and Self-Guided Research	3	10	10	3	6	8

Research themes addressed in the Academic Curriculum session

Educational materials that link junior high and high school curricula

Unit formulation techniques for cultivating students' ability to think on their own

Research on the composition of mathematics sciences that focus on communication

A curriculum that clarifies goals through the use of evaluation standards developed from various perspectives

The formulation of a junior high school mathematics education curriculum

Lesson Study in High Schools

Section 3.1: Current Status of Lesson Study in High Schools

Kazuhiko Murooka

The goal of high school mathematics program is to cultivate a foundation of creativity that allows students to understand and apply mathematical concepts and their systematic relationships. The mathematics curriculum is comprised of core courses that comprise a system leading to precalculus and optional courses to supplement students' development in other areas.

Japanese high schools are ranked according to their entrance exam scores. In some schools, teachers devote a lot of time to teach elementary arithmetic for teaching high school mathematics. The goal of many students at high schools with large populations of college-bound students is to pass the entrance exam at the university they want to attend. Thus, many teaching materials are often selected based on the entrance exams to the college. Even though high schools have problems and diversity, many teachers are engaging in Lesson Study to develop ways of mathematical modeling and to use technology in the classroom regardless of their school's ranking, and the students are learning eagerly.

Classes are most often structured according to the following patterns, and Lesson Study is an extension of these patterns:

(a) teachers explain a typical problem, then drill students,

(b) teachers explain the problem while exchanging questions and answers with students, then drill the students,

(c) teachers address and discuss the students' solutions and then teach new material,

(d) teachers rearrange groups by ability, and adjust what and how they teach based on the students' needs.

Table 1: Major elements of lesson process and Lesson Study (Ago, 2002)

Teaching method	Always	Often	Occasionally	Never
Drills	19%	43%	34%	4%
Discussion	16	33	42	9
Divide class into groups	12	38	31	19
Problem-solving approach	12	32	40	16
Computers	1	5	17	77
Open-ended approach	2	22	49	27

Teaching method	Very good	Pretty good	Not very good	Bad
Drills	6%	71%	21%	2%
Discussion	51	44	5	0
Divide class into groups	21	69	10	0
Problem-solving approach	6	71	21	2
Computers	13	67	17	3
Open-ended approach	31	56	11	2

Table 1 shows the results of a recent survey of high school teachers which recorded their impressions of how often they used specific teaching approaches, and how well they felt these approaches were taught. Even if teachers know that computers and open-ended approach were preferable good ways of teaching, many of them did not choose these approaches in their own classrooms. It seems like self-contradiction or negligence, but they usually have some clear rationality to teach mathematics. Many teachers have their own order for teaching priority based on their experiences and sense of value. A possible explanation of their teaching practice from Table 1 is as follows: Many high school teachers may believe that developing skills by drills are important but time consuming, discussion and communication are important but few opportunities are given for students, problem solving approach looks good but too much time it will take, computer and open approach are attractive ways of teaching but only limited numbers of teachers prefer to use.

The following types of lessons are also being studied:
 (e) lessons that use computers and graph calculators as tools,
 (f) lessons taught using an open-ended approach.
Recently there has been a lot of emphasis in Lesson Study on encouraging basic creativity and mathematical activities. This is done by teaching students to apply mathematics to real-life situations through lessons on such topics as developing mathematical models. Figure 1 points to the interaction between real – life topics or everyday activities and the development of mathematical definitions and theorems. Figure 2 gives an outline of the process of mathematical modeling.

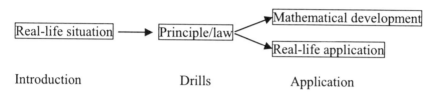

Introduction   Drills   Application

Figure 1: Problem Solving Approach with Drills and Excercise

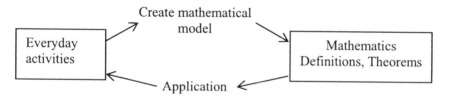

Figure 2: Mathematical Activities

Section 3.2: Changes in the High School Curriculum Based on the National Course of Study

Kazuhiko Murooka

High School Advancement Rate

In 1955, the high school advancement rate was 51%. Ten years later it had risen to 90%, and over the past 20 years, it has risen further to about 95%. During that time, later secondary education has gone from being an education for the elite (in prewar times) to education for all, and schools are now being asked to develop curricula based on diverse academic skills and academic career paths, and to develop practices that can keep up with social change.

As a result, the number of mathematics class hours (number of units) has fallen dramatically. In 1968, students were required to take six hours of compulsory mathematics. That number was reduced to four hours in 1978 and again to three hours in 2003. At the same time, the elective system has become more developed due to the need to maintain the skills of students who want to go on to college.

The diagram opposite shows core courses and optional courses in mathematics in the senior high school in 1978 followed by subsequent adaptations in 1989 and in 2000. Also shown is the number of class teaching hours recommended for each course.

Educational revisions in 1989 established the following core and optional courses:

      Core courses: Mathematics I, II, III

          (focus on functions, precalculus)

      Optional courses: Mathematics A, B, C

          (focus on geometric figures, probability/statistics)

Changes to National Course of Study and Research Topics (continue to page 143)

Major Components of the National Course of Study in 1978

Figures in ( ) indicate the number of class hours

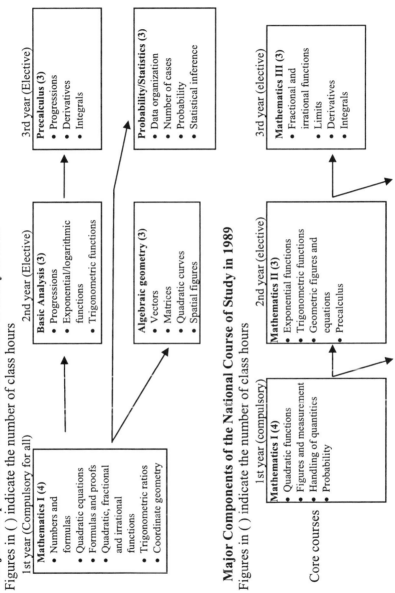

1st year (Compulsory for all)

Mathematics I (4)
• Numbers and formulas
• Quadratic equations
• Formulas and proofs
• Quadratic, fractional and irrational functions
• Trigonometric ratios
• Coordinate geometry

2nd year (Elective)

Basic Analysis (3)
• Progressions
• Exponential/logarithmic functions
• Trigonometric functions

Algebraic geometry (3)
• Vectors
• Matrices
• Quadratic curves
• Spatial figures

3rd year (Elective)

Precalculus (3)
• Progressions
• Derivatives
• Integrals

Probability/Statistics (3)
• Data organization
• Number of cases
• Probability
• Statistical inference

Major Components of the National Course of Study in 1989

Figures in ( ) indicate the number of class hours

1st year (compulsory)

Mathematics I (4)
• Quadratic functions
• Figures and measurement
• Handling of quantities
• Probability

2nd year (elective)

Mathematics II (3)
• Exponential functions
• Trigonometric functions
• Geometric figures and equations
• Precalculus

3rd year (elective)

Mathematics III (3)
• Fractional and irrational functions
• Limits
• Derivatives
• Integrals

Core courses

Courses implemented in 2003 included the newly established "Mathematics Fundamentals," which incorporated materials that addressed everyday phenomena such as statistics, as well as other materials that students were interested in. This made it possible for even the less mathematically skilled students to complete their mandatory three unit hours. However, only 4% of high schools actually offered "Mathematics Fundamentals," leaving most schools to continue teaching the same content to all students, regardless of their skill level.

Japanese schools are supposed to establish their own curricula, but there is a lack of diversity due to the use of officially tested textbooks as primary teaching materials. With the decreasing number of young people leading to reductions in the number of high schools such as the closing and integration of schools, more and more schools are trying to make their curricula more useful in an effort to attract and keep students. Proposals are also being made to develop mathematics curricula that high school students will find interesting so that high school mathematics programs can contribute to this process.

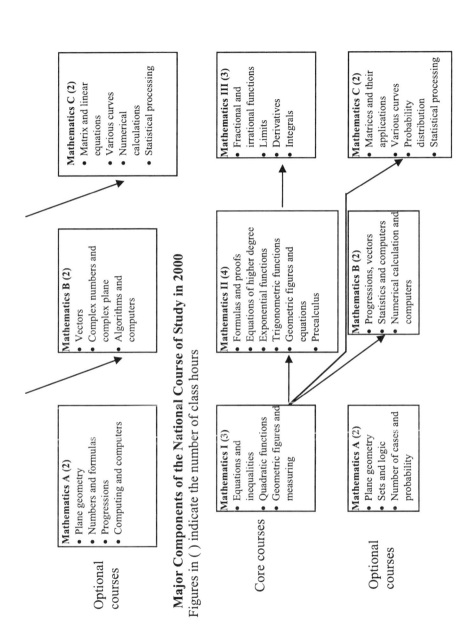

Major Components of the National Course of Study in 2000
Figures in ( ) indicate the number of class hours

Optional courses

Mathematics A (2)
- Plane geometry
- Numbers and formulas
- Progressions
- Computing and computers

Mathematics B (2)
- Vectors
- Complex numbers and complex plane
- Algorithms and computers

Mathematics C (2)
- Matrix and linear equations
- Various curves
- Numerical calculations
- Statistical processing

Core courses

Mathematics I (3)
- Equations and inequalities
- Quadratic functions
- Geometric figures and measuring

Mathematics II (4)
- Formulas and proofs
- Equations of higher degree
- Exponential functions
- Trigonometric functions
- Geometric figures and equations
- Precalculus

Mathematics III (3)
- Fractional and irrational functions
- Limits
- Derivatives
- Integrals

Optional courses

Mathematics A (2)
- Plane geometry
- Sets and logic
- Number of cases and probability

Mathematics B (2)
- Progressions, vectors
- Statistics and computers
- Numerical calculation and computers

Mathematics C (2)
- Matrices and their applications
- Various curves
- Probability distribution
- Statistical processing

Section 3.3: Research Trends in the High School at the Japan Society of Mathematical Education (JSME) National Conference

Kazuhiko Murooka

JSME National Conference Sessions

The JSME National Conference is a venue where participants announce and discuss research on mathematics education practices from elementary school to the university level. In 2002, thirteen sessions were created, and one to two classrooms were used for the presentations in each session over the course of two days. The following topics made up these thirteen sessions:

> General Curriculum relating to senior high school
> General education/science and mathematics courses
> Industry/commerce/agriculture
> Mathematics I, II, III
> Mathematics A, B, C
> Teaching methods and assessment
> Computers and other educational tools
> University entrance examinations
> Basic/independent research

In preparation for the new curriculum being implemented in all grade levels in 2003, sessions on mathematics fundamentals and integrated learning were also added.

Table 1: Number of Presentations in Each Session at the National Conference

Session topic	1996	1997	1998	1999	2000	2001	2002
Education curriculum	18	9	8	11	9	13	14
Mathematics I / A	27	18	12	23	8	17	7
Mathematics II / B	11	11	12	14	4	5	8
Mathematics III / C	4	6	3	8	3	8	3
Computers	23	36	8	14	16	16	13
Teaching methods/evaluations	36	27	22	16	12	18	7
Independent/basic research	27	16	16	22	16	32	21

Trends in the Number of Research Presentations Made at the National Conference

Held concurrently with ICME 9 and for a shorter length of time, the Chiba conference in 2000 was an unusual conference. In most years, the conference features many presentations on computers, teaching methods, and basic/independent research. Pure mathematics sessions may also include presentations on materials study, but in most years there tend to be a lot of sessions on the development of new teaching methods and creative techniques that can be applied in everyday teaching situations. The curriculum and standards used by each school are different, so the announced findings are also unique to specific situations.

There is a great need for research on the educational components required by the new education curriculum, such as "mathematics fundamentals" and the "integrated learning period," so the number of research presentations on these topics is increasing. On the other hand, in spite of the fact that computers are generally not being utilized for actual teaching at the high school level, this field has attracted a great deal of interest and an extremely large number of related presentations. The number of presentations has actually fallen as ICT has come into more widespread use. This may be related to the fact that the new curriculum separates information and communication technology courses from mathematics courses.

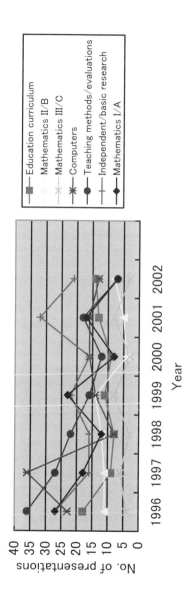

Trends in the Number of Presentations

Themes of the Education Curriculum Session at the 2003 Aichi Conference
History of Japanese mathematics education in the late 20th century
Perceptions of and attitudes toward mathematics
Old and new mathematics curricula at industrial high schools in Aichi Prefecture and Gifu Prefecture
Implementation of the new education curriculum in Osaka's municipal high schools
Implementation of the new National Course of Study
Problems with the implementation of mathematics in integrated classes
Challenges in mathematics education and responses to those challenges
Sample mathematics topic tests
Comparative study of high school students from the perspective of the decline in academic skills

CHAPTER 4

Diversity and Variety of Lesson Study

Case 1: Lesson Study as In-School Training

Hidenori Tanaka

Evidence of effective classroom practice is proved only in the classroom. This is a basic philosophy of Lesson Study. The aims of each school are implemented in various ways, especially in each lesson because 80 percent of pupils' life in school is occupied by lessons on academic subjects: This is a basic philosophy of school management and curriculum implementation. For implementing the aim of the school, we usually engage in following activities:

1. Setting a developmental research theme with 'why?'
   The research theme, a kind of idealistic developed pupil's image by education, is set on socio-cultural and educational necessity based on current pupils' situation (by 'why?').
2. Setting tasks for the theme with 'what' and 'how'
   The tasks for implementation of the theme are usually considered from two perspectives: one of them is related to the subject matter (by 'what?') and another is related to the teaching-learning process (by 'how?').
3. Lesson Study as a method of research and teacher training in school

All teachers should be educational researchers on their own practice such as subject matter and methods of teaching, and evaluating their capacity for developing pupils. Teachers do not necessarily write scientific papers because their research is observed and demonstrated in their study lessons by other teachers. In this sense, Lesson Study is a strong part of an in-school teacher training system. In each lesson, a teacher is expected to learn from pupils and pupils are expected to learn from each other and from the teacher. Thus, Lesson Study is a method by which both pupils and teachers can grow up together with demonstration.

Lesson Study is planned throughout the year in each school. In Japan, each school's principal has an obligation to set a research and teacher training plan. In the case of secondary schools and

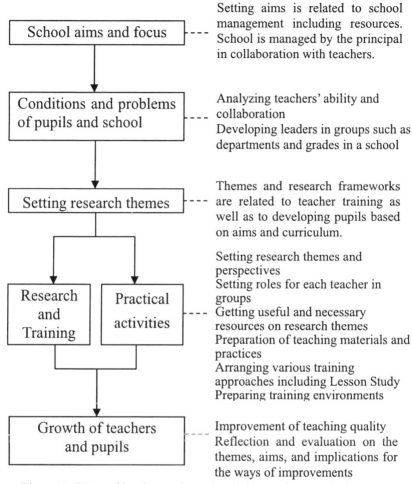

Figure 1: Ways of implementing school aims and research themes

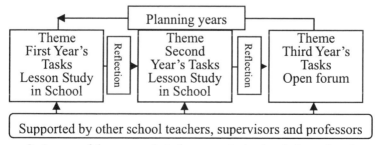

Figure 2: A case of three years' study on central schools in regional area.

special elementary schools such as attached elementary schools, there are departments on academic subjects that plan for a year and Lesson Study is usually managed by each academic department.

In other elementary schools, Lesson Study is usually managed by a research department across the whole school if there are no strong specialists in specific academic subjects. Learning from each other is considered the best way for teachers to share ideas for developing good practice.

For sharing ideas, it is considered better to teach the same or similar content. Thus, Lesson Study groups in each elementary school are usually divided depending on grades and a few selected subjects. In many cases, Lesson Study groups are divided by selected academic subjects according to three age levels (1st & 2nd grades, 3rd & 4th grades, and 5th & 6th grades).

Lesson Study is guided, in a sense, by a research hypothesis linking a research theme (results) to its various implementing tasks (causes). Lesson plans for one hour are developed several times over with a number of teachers working in groups. All group members implement their own study lesson in the course of a year. Teachers usually show their lesson to parents at least once a semester. Study lessons are also open to other teachers in the school. A number of times in a year, a supervisor from a municipal board of education provides comments to all participants in a final reflection session. In some cases, teachers from outside schools are invited to attend the whole process of Lesson Study. In many cases, there are public forums for study lessons and reflections. These provide opportunities for showcasing achieved lessons and for showing what students can achieve as a result of Lesson Study undertaken over several years by a school.

In Japan, differences in pupils' development among classrooms in a school are usually attributed to differences in teaching. For sharing good practice to develop their pupils' capacities, Japanese teachers openly exchange ideas about subject matter knowledge together with detailed knowledge relating not only to what has been taught but also to why and how.

Table 1: A case of year research plan in the case of ordinary school

1st semester	
April	Research framework of the year is presented by the research department. Group meetings for setting research contents.
May	Group meetings for planning the year's schedule.
June	1st & 2nd grade group meetings for planning the Study Lesson. Study Lessons and discussions of 1st & 2nd grade groups. Training on painting by an able teacher in the school for preparing for school painting event.
July	5th & 6th grade group meetings for planning the Study Lesson. Study Lessons and discussions of 5th & 6th grade groups.
August	3rd & 4th grade group meetings for planning the Study Lesson.
September	Study Lessons and discussions of 3rd & 4th grade groups.
2nd semester	
October	School meeting for reflecting on the 1st semester and planning the topics for the 2nd semester.
November	5th & 6th grade group meetings for planning the Study Lesson. Study Lessons and discussions of 5th & 6th grade groups. Training on sculpture by an able teacher in the school for preparing school exhibition.
December	1st & 2nd grade group meetings for planning the Study Lesson. Study Lessons and discussions of 1st & 2nd grade groups.
January	Training on printing by an excellent teacher in the school.
February	3rd & 4th grade group meetings for planning the Study Lesson. Study Lessons and discussions of 3rd & 4th grade groups.
March	School meeting for reflecting on the year and planning the next year.

Case 2: A Study of a Class in the Training Course for Teachers with Ten Years of Experience

Takaharu Komiya

Different aspects of this training course are described in the table on next page. This outlines a training program for junior-high and secondary school mathematics teachers at the Ibaraki Teacher Training Center (Prefectural Level).

The aim of this training course is to investigate the problems of mathematics teaching from a wide perspective and to deepen teachers' awareness. There are two main content areas in this course: 1) to develop lesson plans which include various ideas for practical classes, for example, adoption of mathematical activities, with these lessons being tried within the course, with participants as students;. 2) to carry out research on a theme, chosen by the participants themselves, aimed at the improvement of teaching quality and utilizing sound teaching methods.

Lesson Study is at the core of both these content areas. The main characteristics of Lesson Study are: 1) to base aims and content on the Course of Study of Mathematics, 2) to be able to implement students' mathematical activities, 3) to develop ideas to foster students' basic mathematical skills, 4) to create lessons which students find enjoyable activity and worthwhile, and 5) to use evaluation for the improvement of teaching.

Supervisors in the Teacher Training Center emphasize Lesson Study as an important vehicle for teachers' professional development. Lesson Study by the teachers of Ten Years Training Program is also carried out with collaboration of supervisors from each municipal board of education.

Table 1: Training Course for teachers of ten years' experience

Evaluation and developing plan for teachers' abilities and aptitudes (at the level of Municipal Boards of Education)			
From April to March, school principals evaluate each teacher's ability in subject matter teaching and student guidance. They make an annual training plan tailored to an individual's ability/aptitude and submit it to Municipal Boards of Education. Municipal Boards of Education determine how to carry out and coordinate these plans.			

Carrying out External (off-school) Training (15 days)			
Teacher Training Center at the prefectural level: 6 days during the school vacation season and 6 days during the school year. Municipal Boards of Education level: 3 days			

Teachers' grounded knowledge	Pupils' Human rights and teachers' service obligations	1day	Teacher Training Center at the prefectural level
	Teaching of students' projects, *Sougotekina Gakusyu.*	1day	
	Improving attributes as a key person in each school	1day	
Subject Matter	Movements and trends of reform, expected ways of teaching specific contents and elaborating lesson plans	2days	
	Setting a research topic appropriate for developing lessons, planning the research, and developing the lesson plans, discussion of the lesson plans, presentation of the results from the implementation of the lessons	3days	
Student guidance	The Child Welfare Law The Juvenile Law	2days	
Working experience	Internship training outside of school	3days	Municipal Boards of Education level
Special topics selected by individual teachers	Moral education, Extracurricular activities, Students' projects, Classroom management, School counseling, and Information Technology Education	2days	Teacher Training Center, at the prefectural level

In-school Training (15 days)
Researching on a theme towards the development of teaching materials for a Lesson Study at school: Participants in the Training course -teachers with 10 years of experience- present their lesson in collaboration of colleagues. Principals, vice-principals and curriculum coordinators provide advice and suggestion. Researching on a theme related to teaching methods and subject matter: Participants in the Training course present the result of their research at the end of the school year. Principals provide guidance and advice. Developing teachers' humanity

Evaluation of the results of the training
Principals evaluate teachers' ability for subject matter teaching and student guidance. They report the result to Municipal Boards of Education. Municipal Boards of Education report to Prefectural Boards of Education.

Case 3: Ties between a University Faculty of Education and Its Attached Schools

Hideki Iwasaki

An attached school is a facility where the teacher training and educational research activities for an affiliated university (faculty of education) are conducted. The Japanese teacher development system is an open system that allows students in any faculty to obtain their teaching certification as long as they earn the number of credits required by the Education Personnel Certification Law. The acceptance of teacher trainees and implementation of educational research are not limited to attached schools. The key feature of attached schools lies in the connection between the attached school and the affiliated university's faculty of education.

Students in the faculty of education at the university, the majority of whom want to become teachers, undergo teacher training at the attached school. The attached school, on the other hand, usually holds public research meetings on a regional or national scale for discussing the results of research on educational practices, and the university assists the activities. Teacher training and educational research meetings are two important annual events that link the university and the attached school.

1. Teacher Training

The Enforcement Ordinance of the Education Personnel Certification Law amended in 2000 requires five credits (five weeks) of teacher training for the certification of elementary and junior high school teachers. Thus, during the peak season, the teachers at the attached school have to devote a great deal of time to providing guidance to trainees, not a few of whom only have the goal of obtaining certification.

Figure 1: Handbook for practice teaching
at junior high/high school (left) and at elementary school (right)

Photo 1: An actual lesson of Lesson Study

Photo 2: Students engage in Lesson Study

The Japanese teacher certification system is an open system, but only a limited number of teachers can be hired. So, it may be an unavoidable unsatisfaction that many students will not be a teachers even if teachers guide them a lot.

From the perspective of the university, teacher training is an opportunity for students to integrate the knowledge, understanding, and skills pertaining to the subjects and teaching practices they have studied at the university. From the perspective of the attached school, however, teacher training is where students can get a realistic feeling for the school environment and where the image of the teacher as a professional can be developed.

Thus, the following are generally expected of trainees:

(1) Performance of daily school duties and attainment of teaching experience.

(2) Acquisition of basic teaching skills, such as materials research, lesson formulation, and preparation of questions for students.

(3) Subjective implementation of curriculum and understanding of changes in students' performance.

(4) Self-cultivation of one's own teaching abilities.

As shown in the picture, the final phase of teacher training culminates in Lesson Studies conducted by the trainees by subject. A university instructor attends to analyze and evaluate the lessons from a different perspective than the attached school training advisor.

Figure 3: Results of joint research between the Faculty of Education and attached schools

Figure 2: Educational Research Meeting Poster

2. Educational Research Meetings

Activities for educational practice research are not limited to the attached schools. Schools designated as research schools by the Ministry of Education, Culture, Sports, Science and Technology (MEXT) also perform these functions, as they use administrative support from both central and local governments to actively disseminate information. The educational practice research typically conducted at the attached schools, therefore, is derived from continuous research development and a long-term joint research structure, and flexible cooperative relationship with the affiliated university. The publication of a regular research journal, as shown in the picture, is a result of the joint research system. In recent years, some universities have issued public invitations for participation in joint research projects conducted between the university and their affiliated attached school. Many of these are done with financial assistance.

An example of the flexible cooperation is the support of the university instructor in the educational research meetings held by the attached school. These meetings generally adopt the format of a large-scale subject-based, organized Lesson Study, with university instructors participating either as joint researchers, guidance counselors, commentators, or lecturers. Every year, thousands of teachers participate in these events to observe the results of the academic ties between the university and the attached school, as well as the results of the long-term collaborative discussions conducted in preparation for the educational research meetings.

Thus far, these events have provided a way for participating teachers to witness a model of Lesson Study as it is conducted between a university and its affiliated attached school and to incorporate what they learn to improve their own daily lessons.

Name of Project	Representative	Contact Person	Grant (yen)
Developing Assessment Standards for Secondary School English Program	Shogo Miura Faculty of Education	Kayoko Yamada Fukuyama Secondary School	40,000
Developing Assessment System for Social Studies	Syuji Katayama Faculty of Education	Tanahashi Kenji Faculty of Education	30,000
Problem Solving in Science Learning	Keito Yamazaki Faculty of Education	Koji Mita Mihara Elementary School	100,000
Integration of Thinking and Skills in Mathematics	Keizo Ueta Faculty of Education	Megumi Yajima Mihara Elementary School	33,000
Process of Understanding in Mathematics Learning	Masataka Koyama Faculty of Education	Toshiyuki Akai Elementary School	88,000
Nutrition Education in Home Economics and Housecraft	Keiko Ito Faculty of Education	Hiroko Ishida Shinonome Elementary School	209,000
Basics of Formative in Art	Kazuro Mine Faculty of Education	Ayaka Kokushei Elementary School	60,000

Figure 4: Joint Projects Between Faculty of Education and Attached Schools in the Case of Hiroshima University

Case 4: Curriculum Development at Attached Schools

Yutaka Oneda

One of the roles played by schools attached with national universities is curriculum development. The junior high school attached with the University of Tsukuba has promoted its own curriculum development research since it was founded. The geometry content of the current academic curriculum standards has contracted due the reduction in the total number of school hours. Thus, research is being promoted on the development of a geometry curriculum based on the two pillars of "cultivating spatial concepts" and "cultivating skills of proof and reasoning." Curriculum and teaching methods are two sides of the same coin, and a review of teaching materials is essential to the improvement of teaching methods.

Efforts are being made to promote the formation of a distinctive curriculum based on reviews of curricular flow and techniques for presenting topics. The curriculum is unique in that it focuses on the two pillars and organically links the content (topics) based on the uniform three-year vertical system. Also, the curriculum is formed by the arrangement of and relationship between topics.

Curricular Flow

The subjects covered in geometry education include two-dimensional figures and space figures. The curriculum touches on related space figures during the study of two-dimensional figures to enable students to manipulate figures in space. Projection of figures and cross-sections of three-dimensional figures are also covered as tools for cultivating spatial concepts.

Necessities in relation to pillars

We need to formulate a geometry curriculum focused on the pillars. For cultivating spatial concepts and skills, we must select or develop subject matters and interesting problems, and plan the necessary sequence for developing students' concepts and skills by the map.

We need communication aimed at "cultivating skills of proof and reasoning." For this, it is important to teach students how to formulate an argument and persuade the other person based on common rules and knowledge. Thus, we must emphasize mutual interaction and communication, and strive to cultivate students' development through these activities.

Example of how a unit plan is visualized by using diagrams of problems, topics and aims

The flow of the unit plan "similar figures"

Definition of Similarity

Reduced and enlarged figures

Draw a reduced or enlarged figure and define the similarity

Polygons in which the ratios of the lengths of the corresponding sides and the corresponding angles are equal

Find the similar figures and define the similarity

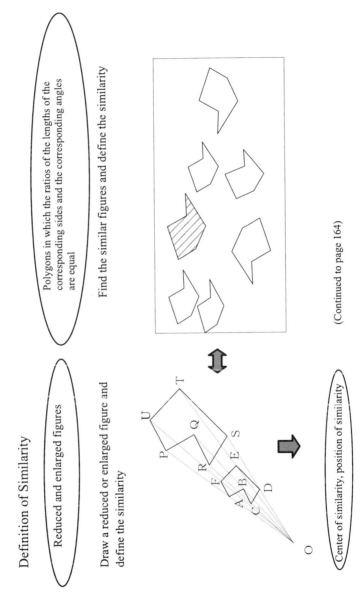

Center of similarity, position of similarity

(Continued to page 164)

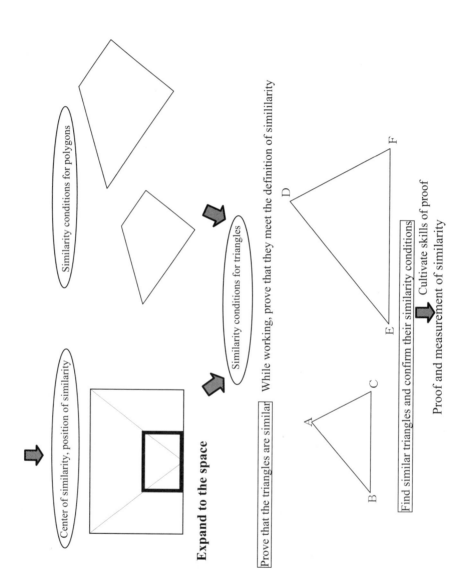

7th Grade (3 hours/week)

Month	Subject
4-5	1. Positive numbers, negative numbers (1) Positive numbers, negative numbers (2) Addition and subtraction (3) Multiplication and division
6-9	2. Symbols and algebraic expressions (1) Symbols and algebraic expressions (2) Operations of algebraic expressions
9-10	3. Equations (1) Equations and their solutions (2) Solving linear equations (3) Applications for linear equations
11-12	4. Quantity changes and proportion (1) Quantity changes (2) Proportions and inverse proportions
12-2	5. Two-dimensional figures (1) Basics of figures (2) Moving figures (3) Figures and constructions
2-3	6. Basics of space figures (1) Space figures (2) Three-dimensional figures and how to investigate their attributes 7. Review of topics learned during the year Curriculum that integrates what has already been learned.

8th Grade A (3 hours/week)

Month	Subject
4-5	1. Algebraic expressions and operations (1) Operations of algebraic expressions (2) Use of algebraic expressions
5-6	2. System of equations (1) System of equations (2) Use of system of equations
7-9	3. Parallelism and congruence (1) Angles and parallel lines (2) Congruence of figures
10-11	4. Triangles and quadrilaterals (1) Triangles (2) Quadrilaterals
12-1	5. Similarity and ratios (1) Similar figures (2) Figures and ratios
2-3	6. Linear functions (1) Linear functions (2) Linear functions and graphs

8th Grade B (1 hour/week) – Latter half of the school year

Month.	Subject
11-3	7. Advanced tasks of space figures (1) Advanced tasks of projected figures (2) Advanced tasks of sections of three-dimensional figures

9th Grade A (2 hours/week)

Month	Subject
4-5	1. Operations of algebraic expressions (1) Multiplication and division of polynomials (2) Factorization (3) Use of algebraic expressions
6-9	2. Quadratic equations (1) Quadratic equations and their solutions (2) Use of quadratic equations
10-11	3. Functions (1) Functions proportional to the square of x (2) Graph of the function $y = ax^2$ (3) Use of functions (Use of technology) (4) Various functions
12	4. Probability (1) Probability (2) Sampling surveys
1-3	5. Review of topics learned during the year Curriculum that integrates what has already been learned.

9th Grade B (2 hours/week)

Month	Subject
4-5	1. Square roots (1) Square roots (2) Square root calculations (3) Rational and irrational numbers
6-10	2. Metric of figures (1) Pythagorean theorem (2) Use of the Pythagorean theorem (3) Cones and spheres (4) Use of similarity (5) Section of three-dimensional figures and metric
11-12	3. Circles (1) Circles and straight lines Chords and tangent lines, circumcenter and inner center (2) Circles and angles Tangent chord theorem and cyclic quadrilaterals (3) Corde and tangent theorem
1-3	4. Review of topics learned during the year Curriculum that integrates what has already been learned.

Figure 1: Attached-School Curriculum (Teaching Plan)

Case 5: Lesson Study: A Partnership among Education Sites, Boards of Education, and Universities

Kazuaki Shimada

How are Lesson Study Partnerships Conducted among Education Sites, Boards of Education, and Universities?

There are three basic partnership formats depending on which institution is taking the leading role in the Lesson Study. These basic types are outlined in the diagram on the opposite page (Figure 1)

First, there is a case in which an education site (i.e. the individual school) plays the leading role for the purpose of improving the practical skills of its own teachers. From the number of lesson studies conducted in Japan, this is the most common type of format, as compared with the two types described below. When the education site is at the center of the study, research may be conducted at the schoolwide level and or conducted primarily by an individual. When schoolwide research is conducted, the research structure centers around the principal, and research topics may be designated by the prefectural or municipal board of education.

Of course, voluntary research may also be undertaken without the external assignment of a research topic. Some research projects being done by individual teachers in partnership with the board of education or a university are conducted using the prefecture's Long-Term Teacher Training System. The partnerships between the three parties are discussed later.

Second, there are cases in which Lesson Study is conducted as part of a research project being undertaken by university professors. In this case, a university professor will form a partnership with the school where the lesson research is to be performed. Sometimes in this case, the research is performed through a relationship with the board of education.

Third, there are cases where a board of education conducts research in partnership with a school site or university for the purpose of improving lessons. For example, the Ministry of Education sponsored an experimental school project at Chiba City Kemigawa Elementary School in the 1940s and 1950s.

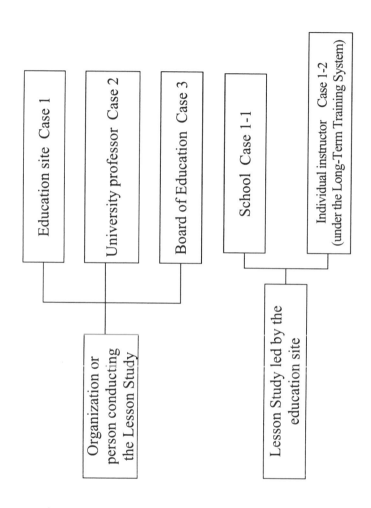

Figure 1: Different partnerships for Lesson Study

Chiba Prefectural Long-Term Teacher Training System.

Under the Chiba Prefectural Long-Term Teacher Training System, elementary, junior high, and senior high school teachers are allowed a one-year paid leave from the classroom to undergo training. In the year from April 2003 to March 2004, about 70 teachers were trained under this system. The majority chose to do their training in the Faculty of Education at Chiba University. Seven of them, four junior high school teachers and three elementary school teachers, underwent training in mathematics. This training program is illustrated by Figure 2.

Research themes proposed by these seven when they began their training were as follows:

- Developing evaluations that enable children to confirm their own learning and voluntarily proceed with their studies.
- To enrich the students' numeral sense and offer mathematics lessons that convey the fun of learning through mathematics activities that emphasize hands-on experience.
- Implementing mathematics education that allows children to learn independently and have fun through unique mathematics activities.
- Teaching methods that improve "mathematical thought processes" through the teaching of geometric proofs.
- Performing evaluations that support the students' independent learning, aiming toward the integration of teaching and evaluation.
- Research on teaching of the notion of function that improves students' mathematical perspective and thought processes
- Teaching methods that firmly implant the fundamentals and basics of geometry through lessons involving the observation and manipulation or hands-on experience with spatial geometry.

These were provisional research topics that were further refined as research progressed. The Lesson Study was conducted by establishing research hypotheses based on these themes and finding support for them based on children's reactions to the lessons implemented.

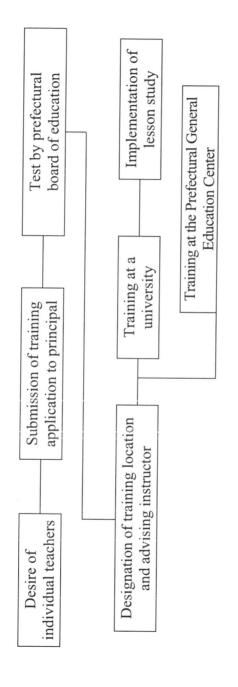

Figure.2: Mathematics Long-Term Training System
(Chiba Prefecture)

Process Leading Up to Lesson Study

Research guidance is primarily provided by the university, but concurrent training by the board of education is also provided. In this way, the process is undertaken in partnership with the board of education. Training sessions by the board of education are held at a prefectural training facility twelve times a year. These sessions cover such topics as how to proceed with research, the development of a research plan, computer use, evaluation methods, compiling research reports, and how to proceed with in-school training.

The list below shows the content of the actual research projects undertaken at the university. The process generally proceeds according to the following schedule:

April: Develop research purpose, review previously conducted researches, create a list of references needed for this research purpose.

May: Develop the research purpose, objectives, and hypothesis. Learn research methods. Study ways to approach references and how to actually implement the research.

June: Proceed with research on the theoretical underpinnings of actual practices. Establish the topic of the study lesson. Create a lesson plan (create unit plan, plan for each class period).

July: Investigate whether research objectives are compatible with the study lesson. Create a lesson plan, diagnostic test (pre-test), and comprehensive test (post-test).

August: Throughout the month, attend research meetings and academic society meetings. In the latter half of the month, develop a study lesson plan for demonstration. Strive to clarify the theoretical components and confirm the demonstration content. Complete the lesson plan.

September: Prepare for lesson implementation. Complete the lesson plan (for the whole unit). Develop a post-test.

October: Implement lesson. Collect and organize data.

November: Clarify information that is emphasized by the data. Create a revised lesson plan.

December: Create a description and straighten the format.

January: Create a list of materials. Compile a report.

A Case

The following describes an actual research that was conducted through a perspective of the partnership among the education site, the board of education, and a university.

The research theme established in April was "Developing Mathematics Evaluations That Enable Children To Confirm Their Own Learning and Voluntarily Proceed with Their Studies Through the Teaching of Decimals."

The research hypothesis was: "Incorporating evaluation standards and criteria in the study of decimals so that students can evaluate their own progress." The anticipated student response as a result of implementing this research was "voluntary efforts in their own learning, the acquisition of self-evaluation skills, and the acquisition of self-learning skills."

At the university, the teachers (researchers) under training identified the theoretical approaches that exist regarding the research objectives and hypothesis as well as their significance, and clarified the position of their own argument within the realm of mathematical theory. The board of education provided guidance with regard to the research framework based on hypothesis testing. Based on this information, the teachers under training made links to specific problems in mathematics and identified elements of the teaching methods that required improvement.

After identifying the problems in mathematics teaching, they moved on to figuring out how to study them in actual lessons. Demonstration lessons in junior high schools are generally conducted in short 3-hour units and in elementary schools in 6-9 hour units. This is possible because trainees are excused from classroom teaching to go off to do research, and thus can conduct a series of Lesson Studies in junior high schools for about three hours and a series of Lesson Studies in elementary schools for about 6-9 hours. Nonetheless, it is rare for a Lesson Study that covers a whole unit to be conducted as an individual's research project.

Materials research has to precede study lessons. Materials research on decimals must be conducted to achieve the research objectives described above. Cultivating self-evaluation skills in children requires the idea to develop children's ideas based on what they already knew and to cultivate children's learning skills for evaluating themselves based on what they learned in relation to what they already knew and the aims of today's lesson. The teaching of decimals taught by the university advisor is an expanding development on the curriculum of integers already learned. That is, after introducing decimals while reviewing the structure of decimal positional notation system with integers, if you apply existing topics under the decimal positional notation system with integers to decimals, the lesson will teach children to ask the question of what can be said about decimals.

When creating the lesson plan, researchers (teachers) generally need to investigate and revise the first hour about five times. This is because they generally create the lesson plan based on the image of the learning activities of the students they taught in the previous year. This results in a variance between the specific teaching activities and the mathematical activities being derived from them. When a university advisor specializing in mathematics education gives a possible mathematical response in response to an activity instructed or directed by the teacher, this will be a more accurate prediction about student behavior than predictions based on the teacher's past experience. Teachers who are accustomed to teaching in front of children cannot predict how children will behave in terms of the consequences of their theoretical ideas when the materials change slightly. University advisors specializing in mathematics and board of education advisors are usually superior resources in this regard.

After creating the lesson plan, the researcher prepares a diagnostic evaluation for ascertaining each student's learning status that can be used by knowing the children's reactions against the hypothesis. That is, to verify changes caused in the lesson, the researcher examines the status of children's knowledge of the mathematics activities to be addressed before the lesson is conducted. Even in this case, the researcher works with the university advisor to develop test problems that have to be addressed from the perspective of the materials and teaching practices.

In the study of decimals, for example, this means the understanding of the decimal positional notation system for integers, the relationship between each position, counting in order, and the composition and decomposition of 10. The researcher creates questions that combine content that covers decimals and content that addresses the research theme, for example, content regarding self-evaluations.

The researcher creates a unit plan and a detailed lesson plan for that period, and then implements the lesson. Trial lessons are often conducted simultaneously in other grade levels in many Lesson Studies. However, there are not many examples in which a lesson plan is created and a Lesson Study conducted for several successive hours of lessons. For university professors, it is not easy to engage in multiple hours Lesson Study with the cooperation of the educators based on the principal's permission. The data corresponding to the pre-tests and post-tests regarding study lesson related to the changes in children's learning can be correlated with the process of the study lessons. The process used to demonstrate the research hypothesis through these procedures is the similar as described elsewhere in this book.

Here, a case of a partnership among an education site, the board of education, and a university is shown. This is an example of how a partnership works when the board of education guarantees the salary and rank of individual teachers and supports the basic components of research, the education site supports the implementation of study lessons, and the university supports major components of the developmental research which is meaningful for the improvement of teaching.

Case 6: Lesson Study Associations

Izumi Nishitani

Lesson Study associations are quite popular in Japan and are making a significant contribution to the improvement of teaching skills among school teachers. Research study associations come in many shapes and sizes, existing at the national level, prefecture level, and local level. In this article I will introduce two Lesson Study associations in which I serve as the chairperson, the Gunma Prefecture Research Society on Mathematics Education and the Mathematics Education Research Association of Gunma University.

1. Gunma Prefecture Research Society on Mathematics Education

This research society is comprised largely of mathematics teachers from all the local elementary, junior high, and senior high schools, as well as professors specializing in mathematics education in the Gunma University Faculty of Education (Figure 1). One of the society's activities is an annual event in which educators make visits to local elementary, junior high, and senior high schools where they conduct research lessons (see Photos 1-4), and engage in mutual training exercises. This allows educators to see how lessons are taught at other schools and provides them with a training opportunity. The society's individual committees also hold Lesson Study sessions. The research findings of the society are published in its bulletin which facilitates information sharing among educators.

2. Mathematics Education Research Association of Gunma University

This association is composed primarily of graduates of the Gunma University Faculty of Education Mathematics Department or Graduate School who have gone on to become school teachers. The association holds monthly meetings (Photo 5) and an annual general meeting (Photo 6). Members gather to discuss topics that focus on their actual classroom experiences, addressing such topics as students' learning problems, lesson devices, and the purposes of and problems involved in teaching. Participants then use what they learn to improve their own classes. The association also publishes a bulletin which announces the papers and the topics being examined.

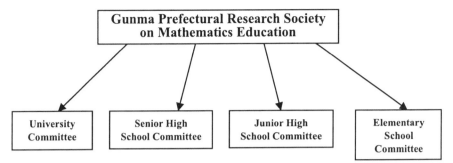

Figure 1: Components of a Lesson Study Association

Photo 1: Research lesson at a
high school

Photo 2: Research lesson at a
junior high school

Photo 3: Research lesson at a
junior high school

Photo 4: Research lesson at a
junior high school

Photo 5: A Lesson Study meeting

Photo 6: A general meeting of
the research group

Case 7: Lesson Study in Teacher Education Programs: How do Students Become Teachers That Implement Lesson Study?

Masami Isoda

Through their experiences of Lesson Study, students learn to think from the perspective of a Lesson Study researcher.

Undergraduate Teacher Education Program: It is difficult to learn to think from the perspective of a teacher even if a student takes classes on a particular academic subject or on materials research. Thus, in teacher education programs, students engage in micro-teaching exercises in which they engage in role playing, alternately playing the role of the teacher and the student to acquire the perspectives of both teacher and learner. They also participate in teaching internships of one month or longer during which they do on-site training in an actual school. This allows students to become familiar with the cyclical Lesson Study process of researching materials, conducting study lessons, and holding feedback meetings to facilitate improvement. In the final week of their teaching internships, students invite their advisor from the university to participate in their own Lesson Study project at the school.

Master's Program: Only a small number of teachers are hired, and those who are tend to be individuals who have become teachers by obtaining their Rank 1 Teaching Certificate in a master's degree program. There are big trends on pre-service teachers going to graduate schools of teacher education, even if

Table 1: University of Tsukuba Master's Program in Education Classes:
Mathematics Pedagogy II and Exercise

Goals of Lesson Study (2002): To develop teaching materials that awaken students' cultural views of mathematics, to track changes in students' attitudes and briefs toward mathematics by conducting lessons using those materials, and to demonstrate the educational value of the materials developed.

(1) *April – June*
Transition period: Second year students conduct classes to review the activities from their actual lessons from the previous year. New first-year students experience these classes from the student's point of view, and also learn how to use the computers used in the Study Lesson.

(2) *July – August*
Reading of sources: Reading of authoritative texts on mathematics history (English translations of primary sources) for excavating teaching materials, reading of *History in Mathematics Education* (Fauvel, J. & Maanen, J., eds., 2000) to learn the educational value and teaching methods of mathematics history.

(3) *September – November*
Materials development: Conceptualize lesson, establish goals, and develop teaching materials using authoritative texts.

(4) *November – December*
Lesson implementation: Teach the lesson.

(5) *December – February*
Report preparation: Write a research report, create a web site

http://www.mathedu-jp.org/forAll/project/history/index.html

they have teacher certificate. Each university's master's degree program offers its own excellent and distinctive teacher's education program. Teacher education programs that cultivate the ability to lead practical and useful educational research are especially welcomed by current teachers, the board of education, and the Ministry of Education, Culture, Sports, Science and Technology.

The Mathematics Course of the University of Tsukuba Master's Program in Education, which aims to train teachers for high school and upper, addresses both pure mathematics and mathematics education by requiring that the master's thesis be comprised of a main thesis and a supplementary paper, and strives to cultivate students who are well rounded in both specialized subject matter and educational topics. In their first year of the two year program, graduate students develop original mathematics teaching materials and conduct a three-hour Lesson Study project in which they teach using those materials and study changes in students' learning. Thus, even if a student strives to achieve original results by writing their master's thesis on a mathematics topic, their Lesson Study skills are ensured by the requirement that they compile their Lesson Study findings in the supplementary paper. And if a student writes their main thesis on an education-focused topic, the Lesson Study requirement ensures his/her ability to write a research paper with practical value backed by actual data from those Lesson Study experiences.

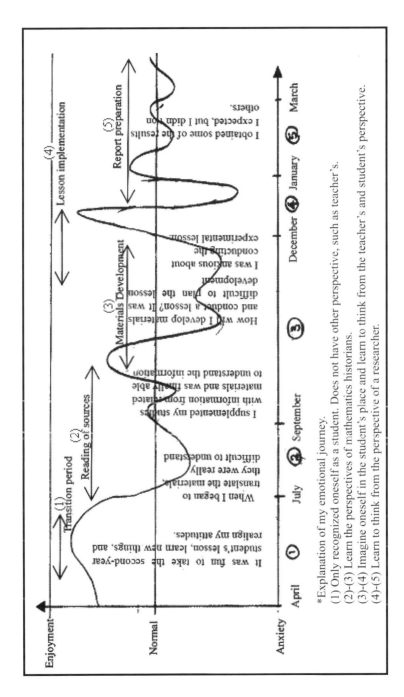

Figure 1: Changes in attitudes reported by one graduate student in the Lesson Study project process (Isoda, 2002)

Case 8: Lesson Study Project Supported by Ministry of Education, Culture, Sports, Science and Technology: How to Effectively Use Computers in Class

Yasuyuki Iijima

How should it be possible for all teachers to use computers in the classroom?

1. "Computerization of Education" and the Project to Promote the Digitization and Network Adaptation of Learning Resources

Japan implements the "Computerization of Education" project (2000-2005), changing its direction of the use of computers in school education. This project aims to introduce computers (with Internet connectivity) and projectors in regular classrooms, making it possible for all teachers to incorporate computers into their everyday lessons. The Project to Promote the Digitization and Network Adaptation of Learning Resources developed content for use in school education. It emphasized the importance of addressing the needs and opinions of teachers, the users of this content, and divided content development into three stages. Developers engaged in the process of Lesson Study, discussion, evaluation, and improvement.

2. Using lessons to verify content effectiveness and acquire how to practice

The image shown at right is a Lesson Study conducted by a dynamic geometry software consortium. Based on these kinds of classes/investigatory meetings, ways of developing content appropriate for the classroom, ways of using tools, and teaching strategies were obtained. Also, a video of the class was sent to teachers who were unable to participate in the Lesson Study meeting, and information about the content was exchanged on a mailing list. The method of projecting images on a blackboard and then using chalk to mark or add symbols or words to the images is a typical example of how applications were derived from the Lesson Studies.

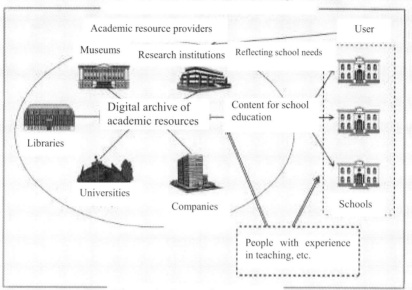

Figure 1: Consortium Image

Brainstorming. Investigation in groups. Discussion

Figure 2: Discussions by the consortium

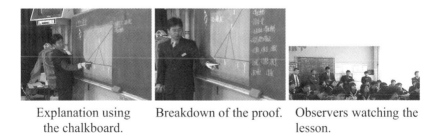

Explanation using Breakdown of the proof. Observers watching the
the chalkboard. lesson.

Figure 3: Lesson Study

CHAPTER 5

International Cooperative Projects
in Education

Case 1: International Comparative Classroom Research Project
What are the Characteristics of Japanese Lessons Emerged by the International Comparisons?

Yoshinori Shimizu

1. Focus of the International Comparison of Lessons

We can learn something about the unique characteristics of lessons, which are both cultural and social in nature, in Japan by comparing them against the practices of other countries. International comparisons are conducted against the backdrop of classroom culture at every level, on the goals of lesson and perspectives on teaching and the topic to be taught described in teaching plans (intended lessons), lessons as they are implemented in the classroom (implemented lessons), and what the students and teachers learn or achieve as a result of lessons (accomplished lessons). Figure 1 summarizes these three levels. The recent international comparative studies on mathematics lessons that include the analysis of videos as a major methodology have revealed several unique characteristics of Japanese mathematics lessons. Figure 2 summarizes those characteristics as identified by the analysis of videos.

2 Results of Large-Scale International Comparative Research on Lessons and Their Impact

The Videotape Classroom Study conducted as part of the Third International Mathematics and Science Study (TIMSS) revealed several unique characteristics of Japanese mathematics lessons by comparing them against practices in other countries. For example,

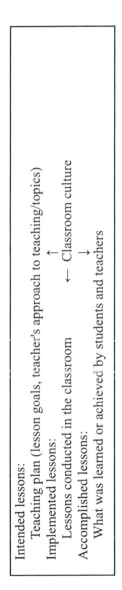

Figure 1: Focal Point of International Comparative Research on Lessons

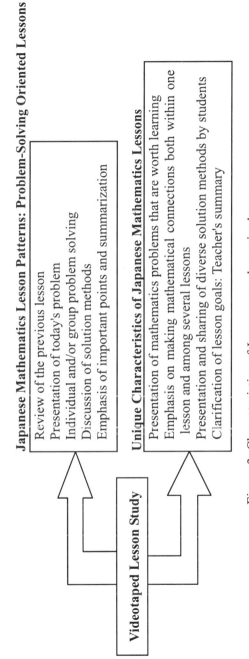

Figure 2: Characteristics of Japanese mathematics lessons

Japanese teachers deepen discussions and make summarizations while emphasizing the students' diverse ideas and thought processes regarding the problems presented (See Figure 3).

Based on these results, a new research project known as a Learner's Perspective Study (LPS) has been conducted to identify the components of the consecutive lessons within a teaching unit, and to clarify the differences of the perceptions of lesson events between the teacher and the learner.

The LPS looks at elements of a lesson obtained from a series of lessons. In the LPS the unit of study is a series of lessons, not a single lesson as in the case of the TIMSS Videotape Classroom Study. The LPS focuses on lesson development and its significance as seen through the eyes of the learner. This is one of the challenges of this study, as an example of international comparative research on lessons.

Meanwhile, the results of the TIMSS: Videotaped Classroom Study have shed new light on Lesson Study activities as sources of lesson "scripts" shared by Japanese teachers, and on the role of research lessons in the development of teachers' skills. These are referred to in Figure 4.

Aspects of each lesson as embedded in a series of lessons (unit)

Lesson development and its significance as seen through the eyes of the learner

$\rightarrow$ Analysis of the lesson as seen through the eyes of the learner

Figure 3: Challenges in International Comparative Research on Lessons

Class observations and intensive discussion of it at Lesson Study meetings

Challenging towards a research lesson   Teaching methods, sharing of perspectives on the topic and students

Figure 4: Sources of Skill Development among Japanese Teachers

For those wishing to obtain more information about these studies and perspectives, following Web Sites can be recommended:
http://www.lessonlab.com/ and http://www.edfac.unimelb.edu.au/DSME/lps/

Case 2: Lesson Study in Thailand

Maitree Inprasitha

1. Introduction

After the 1999 Educational Acts were enacted, Thailand was put into an educational reform movement. Most school teachers have been attempting to improve their teaching practices. Unfortunately, they do not find good ways to improve their everyday work. Most teachers still use a traditional teaching style focusing on coverage of content, but they do not emphasize students' learning processes and their attitudes toward learning with understanding. More importantly, a number of teachers classify themselves into a reforming group (e.g., master teachers and inovative teachers) but, in effect, do not realize that they are still in an old paradigm group.

Lesson Study is a comprehensive and well-articulated process for examining practices that many Japanese teachers are engaging in (Fernandez, Cannon & Chokshi, 2003). In fact, recently a number of American researchers and educators have suggested that Lesson Study might be an incredibly beneficial approach to examining practices for US teachers (Lewis, C., 2002; Fernandez et al., 2003).

In Thailand, there is also another initiative to use Lesson Study to improve the teacher education program in mathematics. In 2002, the Faculty of Education at Khon Kaen University, in an attempt to improve the teacher education program, conducted a project to investigate how student teachers develop their own pedagogy. In addition, to investigate how students in the classroom are responding to the open-approach teaching method and wheather or not they recognize their learning experiences.

2. Initiative Project in Mathematics Teacher Education

2.1 Background of the Project

The research project was conducted in the 2002 academic year in 7 schools in Khon Kaen province in the northeastern part of Thailand. It was aimed at investigating changes in student teachers' pedagogy and their professional development when using the open-approach teaching method (Nohda, 2000). The project was also aimed at clarifying how school students recognize their learning experiences. The following process of Lesson Study has been conducted: cooperatively constructing lesson plans, implementing those plans in the classroom, discussion about lesson plans and individual teacher teaching progression. Fifteen 4th-year student teachers voluntarily participated in this project. According to the requirements of the mathematics education program, they conducted their practice teaching at their selected schools for one semester. They had to follow some regular activities designed by the program and had to follow some additional required activities designed by the research project. In what follows, regular activities and required activities for this project are described.

2.2. The Research Project Settings

2.2.1 Regular Activities Requiring all Student Teachers to be done

All student teachers were required to teach at schools in the Khon Kaen urban area 6-8 periods (about 50 minutes per period) a week. School teachers who serve as school supervisors may assign appropriate work to the student teachers. For one semester, the student teachers were supervised four times by school supervisors and another three times by supervisors from the faculty of Khon Kaen University. They also had to conduct an action reasearch project under the stewardship of his/her research advisor. Furthermore, they had to attend three-hours of seminar and/or to meet with their research advisors once a week.

2.2.2 Required Activities for Student Teachers in the Project

Fifteen student teachers who participated in the research project had attended a one-month workshop for constructing lesson plans to be used later in the first semester of 2002 academic year. They were grouped according to school levels they intended to teach. Six were in the 7th-grade group. Five were in the 8th-grade group and four in the 9th-grade group. Coached by the researcher, they spent about 6 hours a day constructing lesson plans using open-ended problems.

In order to dialouge and review their experiences of the open-approach teaching method, the 15 student teachers attended a special seminar organized by the researcher weekly. In this seminar they expressed their common concern, interesting points and change in some particular students' behavior. Furthermore, they were expected to develop ideas for the conduction of their action research projects. They also kept a journal during the semester related to their teaching experiences. This journal was used for discussion in the special weekly seminar and for data analysis of the research project.

2.3 Research Results

2.3.1 Change in Student Teachers' Pedagogical Practices

During the first half of the semester all student teachers in the project experienced some difficulty adjusting to their new teaching roles and classroom organization. Participation in the weekly seminar facilitated the student teachers gradual change of the teachers' role. The most critical point of change was encountered while sharing their differing teaching experiences among friends and colleagues. Sharing experiences with their friends during the weekly seminar not only resolved their common concerns but also developed and expanded their own pedagogy, teaching practices and professional development. The greatest paradigm shift for student teachers' was that teaching mathematics does not mean focusing on the coverage of content but emphasizing the students' learning processes, original ideas, attitudes towards learning mathematics and satisfying one's own competence.

Most of the student teachers saw the positive benefits of conducting action research while simultaneously completing their student teaching. They have come to realize that doing action research can help them develop a wider perspective on how to view their classrooms. Moreover, they acknowledged that action research may help improve teachers' everyday practices. Most importantly, student teachers in the project changed their attitudes towards learning from academic learning to life-long learning. Their thought process on teaching and learning has been shifted into a new one which is seen as a unification of living and learning. This also influences their values of their own contribution to society, the core values we need for Thai society.

2.3.2 Experiences of School Students on Learning through the Open-Approach Method

According to the survey results of 1200 students in all schools in the project, most of the school students have positive attitudes towards learning through the open-approach method. In all areas of the survey, the school students indicated a marked improvement in their learning environment and capabilities in comparison to their traditional classroom. Regarding the classroom activity, the school students responded that they have more opportunities to act, think, play an active role, do something original, and conclude things by oneself. Regarding the change of their own learning process, they show some interesting responses as follows: more reasonable, more skillful in observation, more cool-hearted, know how to work cooperatively and more confidence in asking "why?" and "how come?" type questions.

2.3.3 Expansion of Lesson Study Approach

The project provides many ideas for the implementation of the Lesson Study approach and on-going professional development. Student teachers in classroom settings provides for an invaluable opportunity to further educate the professional teacher. It is worthwhile to conceive that programs for professional development should start in the earlier years of teacher education programs and be continuous for seasoned professional teachers. So far, Lesson Study approach has provided great influence on the reform of professional development in Thailand. The National Commission on Science and Mathematics Education incorporates the concept of Lesson Study into the frameworks on the development of science and mathematics education. In 2004, the Office of Basic Education Commission provided funding support to organize training for supervisors in order to supervise school teachers participating in the Lesson Study project of Khon Kaen University.

On the international level, Khon Kaen University in cooperation with Minsai Center at Laos PDR, East Asian Circle of Applied Technology of Japan, and Educational Development Fund have organized training programs for mathematics and science teachers from Laos PDR since 2002. This training program also implements the integrated open-approach teaching method and Lesson Study approach. In 2004-2005, Khon Kaen University cooperated with Plan International Organization which also implemented the Lesson Study approach to improve mathematics teaching in the northeastern part of Thailand. This kind of professional development has created teacher networking among countries in the Great Mekhong Sub-region.

References

Lewis, C. (2002). *Lesson Study: A Handbook of Teacher-Led Instructional Change.* Philadelphia: Research for Better Schools, Inc.

Fernandez, C., Cannon, J., & Chokshi, S. (2003). A US-Japan Lesson Study collaboration reveals critical lenses for examining practice. *Teaching and Teacher Education. 19.* 171-185.

Nohda, N. (2000). *A Study of "Open-Approach" Method in School Mathematics Teaching.* Paper presented at the 10th ICME conference, Makuhari, Japan.

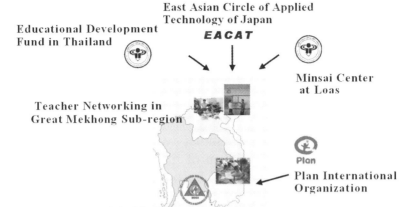

Figure 1

Photo 1

Photo 2

Case 3: Lesson Study in North America

Akihiko Takahashi

In the U.S.A. and Canada today, there is an active movement in mathematics teaching that is striving to improve teaching using the Japanese model of Lesson Study. This movement was inspired by *The Teaching Gap* (J. Stigler & J. Hiebert, 1999), an easy-to-understand compilation of the results of a comparison of junior high school mathematics lessons in Japan, Germany, and the U.S.A.. The book argues that Japanese mathematics lessons incorporate the results of educational research more than those of the other countries and are the closest to the ideal that is being sought in the field of mathematics education. It also draws attention to Japanese-style problem-solving oriented lessons. Having learned about the role that Lesson Study plays in the process of launching Japanese-style problem-solving oriented lessons, schools and school districts (boards of education) throughout the U.S.A and Canada have begun to try implementing Lesson Study practices.

According to the Columbia University Lesson Study Research Group (directed by Clea Fernandez), there are 140 Lesson Study groups active in 29 U.S.A. states, and more than 1,100 educators from 245 schools in 80 school districts are involved in Lesson Study (as of September 2003). And the movement seems to be spreading across North America, as the first public Lesson Study event was held in the Canadian capital of Ottawa in 2004.

What is most important in this stage of the process is that high-quality Lesson Study is conducted, and that many teachers are given opportunities to learn exactly what Lesson Study is. It is essential that meetings do more than discuss surface-level matters or fall back on the notion that "classroom observation alone is enough." Instead, participants have to discuss specific issues, such as where to identify the value of their teaching materials, what should be the role of the teacher, and what improvements need to be made to create better teaching plans.

Photo 1:
About 80 participants who attended a public Lesson Study event held in December 2004 in the Canadian capital Ottawa had a chance to experience a research lesson and a feedback session in a temporary classroom set up in a junior high school gymnasium.

References

Stigler, J. & Hiebert, J. (1999). *The Teaching Gap: Best Ideas from the World's Teachers for Improving Education in the Classroom.* New York: Free Press.

Takahashi, A. (2000). Current Trends and Issues in Lesson Study in Japan and the United States, *Journal of Japan Society of Mathematical Education*, Volume 82, Number 12: 49-6, pp.15-21.

Takahashi, A. & Yoshida, M. (2004). How Can We Start Lesson Study?: Ideas for establishing Lesson Study communities. *Teaching Children Mathematics*, Volume 10, Number 9. pp.436-443.

Yoshida, M. (1999). Lesson Study: A case study of a Japanese approach to improving instruction through school-based teacher development. Dissertation, Department of Education, University of Chicago.

Yoshida, M. (2001). American educators' interest and hopes for Lesson Study (jugyokenkyu) in the U.S.A.. and what it means for teachers in Japan. *Journal of Japan Society of Mathematical Education*, Volume 83, Number 4: 24-34.

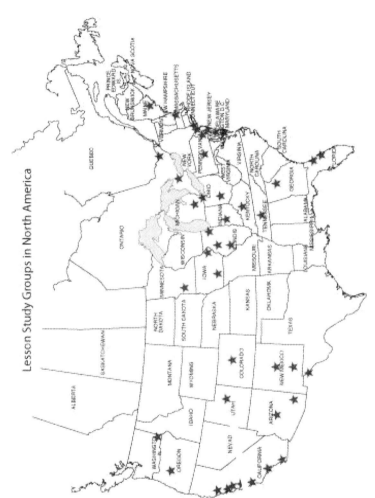

Figure 1: Map of Lesson Study groups in North America

Case 4: Lesson Study for the Effective Use of Open-Ended Problems

Yoshihiko Hashimoto

"Solutions to mathematics problems are either right or wrong; there is only one correct answer." Problems set up in this format are called "closed-ended problems," and stand in contrast to "open-ended problems," or conditional problems in which several correct answers are possible, that is, problems where the result is, as indicated by their name, open-ended. These were initially devised for evaluating high level objectives of mathematics education about 30 years ago (Becker, Jerry P. and Shigeru Shimada, eds. *The Open-Ended Approach: A New Proposal for Teaching Mathematics.*[1]).

In 2002, The Ministry of Education, Culture, Sports, Science and Technology (MEXT) published *Teaching Materials for Individually Adaptable Teaching: Promoting Developmental and Supplemental Learning (for Elementary School Mathematics)* in Japanese, which institutionalized the use of open-ended problems.

The three elements of Lesson Study are (1) the lesson plan (researching material), (2) lesson observation (study lesson), and (3) the feedback session. Consider, for example, a lesson in the last unit on the multiplication table in which students are asked to "find patterns in the multiplication table." This example can be used from elementary school to high school, but the actual wording above was taken from a high school example.

---

[1] Translated from the 1977 Japanese version by S. Shimada, ed., National Council of Teachers of Mathematics, 1997

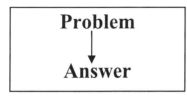

Closed-ended problem

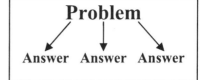

Open-ended problem

1	2	3	4	5	6	7	8	9	10
2	4	6	8	10	12	14	16	18	20
3	6	9	12	15	18	21	24	27	30
4	8	12	16	20	24	28	32	36	40
5	10	15	20	25	30	35	40	45	50
6	12	18	24	30	36	42	48	54	60
7	14	21	28	35	42	49	56	63	70
8	16	24	32	40	48	56	64	72	80
9	18	27	36	45	54	63	72	81	90
10	20	30	40	50	60	70	80	90	100

Multiplication Table

Instructions:
Look closely at how this table is aligned and identify as many properties of the table as you can.

To draft a lesson plan (element 1), the teacher has to try to think of as many of the possible student responses to an open-ended problem, "which has many correct solutions," as possible. After conducting the lesson observation (element 2), it is important that the "many correct answers" be organized in the feedback session (element 3). Sharing the findings of the Lesson Study and working with people overseas on joint research projects are valuable ways in which teachers can learn from one another.

After the 9th International Congress on Mathematical Education (ICME-9) in 2000, post-ICME-9 seminars (between Japan and the USA) were held in Japan at the National Institute for Educational Research and the University of Tsukuba Attached Elementary School. The seminar focused on elementary school Lesson Study. Lesson Study events were also held at elementary, junior high, and senior high schools in 2001 (in North Carolina) and 2002 (Kanagawa Prefecture).

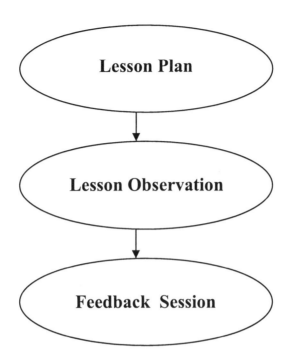

The three elements of Lesson Study

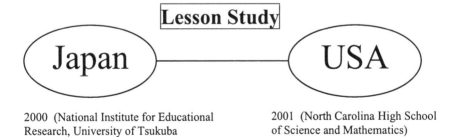

2000 (National Institute for Educational
Research, University of Tsukuba
Attached Elementary School)

2001 (North Carolina High School
of Science and Mathematics)

2002 (Yokohama National University)

Case 5: Lesson Study in Philippines Science Teacher Training Center (ISMED-STTC) Project in the Philippines: How Can the Quality of Education in Developing Countries Be Improved?

Shizumi Shimizu

Project Overview

The Science Teacher Training Center (STTC) Project was developed by the Japan International Cooperation Agency (JICA) in a 5-year plan launched in 1994. With the construction of the center to be funded by the Official Development Assistance (ODA), the project aimed to promote technology transfer and cooperation, and to contribute to the improvement and development of science and mathematics education in the Philippines. The center developed a teaching plan/teaching manual for elementary school mathematics (ESM) and high school mathematics (HSM) teachers, and implemented training in teaching methods and materials development. Technology transfer was promoted by having dispatched specialists transfer technologies to the ISMED-STTC[1] staff and conducting a national training program (NTP), as well as by sending counter part trainees to Japan for training. This cooperative education development program centered around the ISMED-STTC has been highly regarded as a model for how to provide cooperation to developing countries, and has spurred the expansion of projects currently being developed in several other countries.

---

[1] Institute for Science and Mathematics Education Development (ISMED) –STTC was established with support of JICA, and currently, it is National Institute for Science and Mathematics Education Development in University of the Philippines (UP-NISMED).

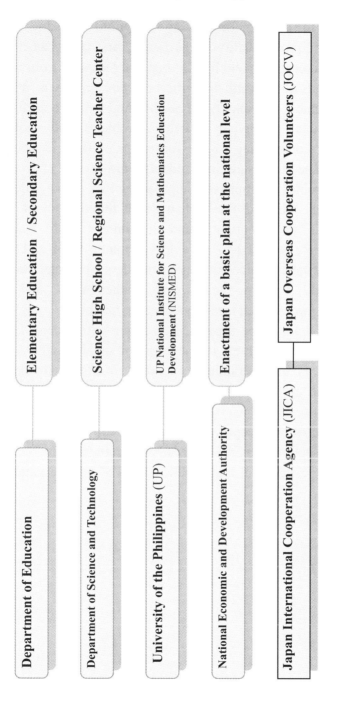

Figure 1: Basic scheme of the Philippines Project

Suggestions for Further Improving the Quality of Education

To further improve quality in the future, first there has to be a shift in the perceptions held by children and teachers, from seeing mathematics as a "facts and drill-oriented" discipline to a "process and idea-oriented" discipline. We need to highlight the usefulness of mathematics in solving everyday problems and show its relevance in contributing to the formation of thought patterns. Second, we need to ensure that students learn diverse approaches to solving problems, emphasize approaches that utilize inductive and functional thinking, establish ways of thinking about "why process skills are important," and disseminate those ways of thinking. Third, we need to develop systematic ESM and HSM curricula that take into consideration the developmental stage of the students, systems of pure mathematics, and the interrelationship between the subjects being taught.

Regarding above three aspects, it is important to consider the appreciation, the usefulness or the value of subject matter in the process and on the curriculum. In order to enable students to recognize the appreciation of mathematical ideas and ways of mathematical thinking, we must consider why and how we teach subject matter more deeply. The improvement of the quality of education is supported by the knowledge of contents which clearly objectify the pedagogical-mathematical aims of subject matter.

Phase I (1986-1991):

Establishment of the Science Teacher Training Center (UP-ISMED-STTC)

Phase II (1992-March 1994 : June 1994-1999)

Preparation and development of SMEMDP (Science and Mathematics Education Manpower Development)

Phase III (June 1999 -   )

Follow-up (School Based Training Program (SBTP): 1999-2001; Project for Strengthening the School Based Training Program for Elementary and Secondary Science and Mathematics Teachers (SBTP-ELSSMAT): February 2002-February 2005)

Figure 2: History of the Philippines Project

Organization names in Figures 1 and 2 were current as of the project launch.

Case 6: Lesson Study in Cambodia
How to Develop Mathematics Teachers Who Can Answer Students' Questions?

Kenji Odani

Attaining credible academic skills by solving problems that require mathematical thinking

1. The Situation in Cambodia

Cambodia's unique educational problem is its lack of trained professional people. From 1975 to 1979, during the rule of Pol Pot, many teachers, from the elementary school to the university level, were killed. Many books were also burned. As a result, the proportion of people with a comprehensive secondary school education, including secondary school teachers, has suffered compared to those in surrounding countries. The government of Cambodia has taken serious steps to correct this situation.

2. STEPSAM Activities

The Secondary School Teacher Training Project in Science and Mathematics in Cambodia (STEPSAM) is an educational support program launched by JICA to help improve this situation. STEPSAM aims to improve the skills of secondary school teachers and is centered around the Faculty of Pedagogy (FOP), Cambodia's only training institution for secondary school teachers. Activities in the mathematical sciences are conducted as shown in Figure 1 and Photo 1. To improve the skills of FOP teachers, training was provided in Japanese universities and intensive summer and fall courses were implemented. FOP teachers have actively served as instructors in the eight rounds of on-site teacher training sessions that have been conducted in Cambodia thus far.

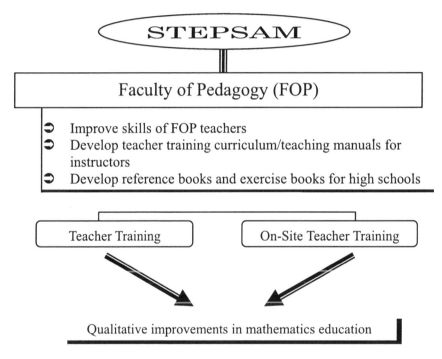

Figure 1: Mathematics Department Activities Under STEPSAM

3. Imparting Credible Academic Skills

The mathematics lessons that the people of Cambodia experienced in the past were conducted under a knowledge-injection style of teaching in which they simply memorized solutions. Thus, around the time that STEPSAM was launched, there was hardly anyone who could solve new problems by exercising their own capabilities. However, STEPSAM has conducted repeated exercises in solving problems that require mathematical thinking. As a result, a pool of people who are able to solve new problems by exercising their own capabilities is gradually being developed.

(Photo courtesy of Koji Takahashi)

Photo 1: On-site teacher training session in outlying areas.

Case 7: Lesson Study in Laos
How to Raise the Quality of Elementary and Junior High School Mathematics Education to International Level?

Noboru Saito

To improve the teaching skills and to enhance the special knowledge on mathematics education of teachers in Teacher Training Colleges in Laos, and to disseminate those improvements throuhout Laos

1. Training in Japan for Teachers in Teacher Training Colleges from Laos

To improve the quality of mathematics education in Laos, cooperative efforts have been underway for the past six years to revise and create school textbooks and to create teaching guidebooks for teachers. In continuation of the cooperation provided thus far, a new JICA project was launched in 2004 (Figure 1) to enable one-fourth of all teachers in Teacher Training Colleges in Laos to spend two months participating in the following Lesson Studies on elementary and junior high school education in Japan over the next four years.

a) The structural learning elements analysis of the teaching contents of elementary and junior high school mathematics textbooks (Photo 1).
b) The new teaching methods that activate creative thinking of students. (Mountain-Climbing Learning Methods)
c) The methods of creating lesson plans and conducting lesson evaluations to achieve teaching goals.

(1) Training in Japan for Teachers in Teacher Training Colleges from Laos

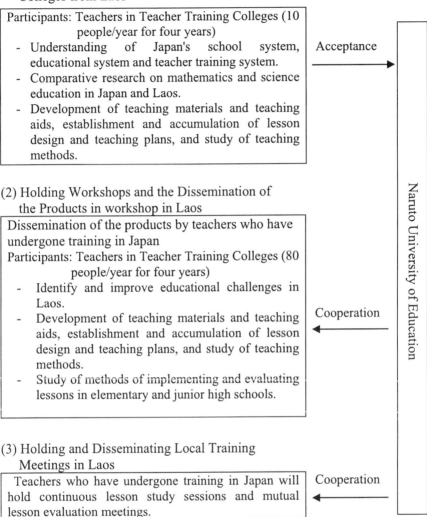

Participants: Teachers in Teacher Training Colleges (10 people/year for four years)
- Understanding of Japan's school system, educational system and teacher training system.
- Comparative research on mathematics and science education in Japan and Laos.
- Development of teaching materials and teaching aids, establishment and accumulation of lesson design and teaching plans, and study of teaching methods.

Acceptance

(2) Holding Workshops and the Dissemination of the Products in workshop in Laos

Dissemination of the products by teachers who have undergone training in Japan
Participants: Teachers in Teacher Training Colleges (80 people/year for four years)
- Identify and improve educational challenges in Laos.
- Development of teaching materials and teaching aids, establishment and accumulation of lesson design and teaching plans, and study of teaching methods.
- Study of methods of implementing and evaluating lessons in elementary and junior high schools.

Cooperation

(3) Holding and Disseminating Local Training Meetings in Laos

Teachers who have undergone training in Japan will hold continuous lesson study sessions and mutual lesson evaluation meetings.

Cooperation

Naruto University of Education

Figure 1: Overview of the Laos Elementary and Junior High School Education Project (JICA)

d) The development of teaching materials and teaching aids, and the practice class that utilize these resources.

2. Holding Workshops and the Dissemination of the Products in Laos

The teachers who have undergone training in Japan will take a leadership role in holding workshops in Laos and provide instruction in Lesson Study methods (Photo 2). The teachers in Teacher Training Colleges who participated in this training program will improve their teaching skills by conducting practice classes in elementary and junior high schools and enhance their teaching ability.

3. Holding and Disseminating Local Training Meetings in Laos

The teachers who have undergone training in Japan will disseminate what they have learned to teachers throughout Laos by holding Lesson Study meetings and lesson evaluation meetings several times year in various regions.

Photo 1: Structural Learning Elements Analysis of the Contents of
Textbooks

Photo 2: A workshop in Laos

Case 8: Lesson Study in Indonesia
How Can Traditional Lessons Be Improved?

Kiyoshi Koseki

1. What is the purpose of JICA's Indonesia Mathematics Education Improvement Project?

The purpose of the project is to "improve the mathematics skills of elementary, junior high, and senior high school students in Indonesia." This is being approached in two ways:

(1) Develop the curriculum of the faculties of education that train elementary, junior high, and senior high school teachers

(2) Develop elementary, junior high, and senior high school lessons

2. How are mathematics lessons conducted in Indonesia?

The teaching methods used by most elementary, junior high, and senior high school teachers have relied on the so-called traditional methods: the "copy method" and the "lecture method." Even in problem-solving learning lessons, the teacher would give the students a problem, have the students think about the answer, and then, instead of walking around the classroom and see how the students worked out their answers, would simply announce the answer after a certain period of time. During this time, the students copied the problem written on the chalkboard into their notebooks. This does not allow the students to appreciate various methods of solving problems.

3. Are improvements being made to teaching methods?

Traditional teaching methods are effective at imparting "skill proficiency" in areas like computation, but a mathematics curriculum also strives to teach an understanding of abstract mathematical concepts and problem-solving skills. Trying to teach all of these things using the traditional method alone has been pointed out to be a major problem recently by university scholars, researchers, and teaching advisors and improvements are now gradually being implemented.

4. What kind of research is being done to facilitate lesson improvements

To promote lesson improvements at schools, university instructors, researchers and schoolteachers are now working together on the School and University Experience Exchange to develop "Pilot lessons" and "Action Research." The recent shift away from teacher-led lessons toward student-centered lessons is one of the results of these research efforts.

Flow Chart of the "Pilot" Lesson Study

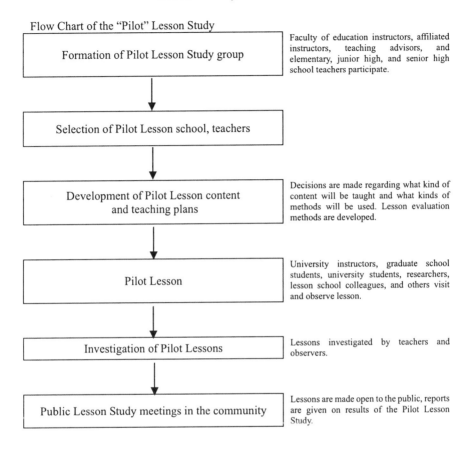

Formation of Pilot Lesson Study group	Faculty of education instructors, affiliated instructors, teaching advisors, and elementary, junior high, and senior high school teachers participate.
↓	
Selection of Pilot Lesson school, teachers	
↓	
Development of Pilot Lesson content and teaching plans	Decisions are made regarding what kind of content will be taught and what kinds of methods will be used. Lesson evaluation methods are developed.
↓	
Pilot Lesson	University instructors, graduate school students, university students, researchers, lesson school colleagues, and others visit and observe lesson.
↓	
Investigation of Pilot Lessons	Lessons investigated by teachers and observers.
↓	
Public Lesson Study meetings in the community	Lessons are made open to the public, reports are given on results of the Pilot Lesson Study.

Photo 1   Photo 2

Case 9: Lesson Study in Egypt

Kazuyoshi Okubo

1. Mathematics Education in Egypt

The Hokkaido University of Education, through a partnership with JICA, is involved in the Elementary School Mathematics Education Improvement Project in Egypt. A guidebook (GB) for teachers was created as part of a three-year mini-project from 1997-2000. The goal of the project for 2003-2006 is to further develop this GB, use it to conduct actual lessons in four pilot schools, verify and revise it, and then disseminate it through research lessons.

Mathematics education in Egypt is generally characterized by teacher-led "instruction" which is then followed by practice. Also, textbooks offer little systematization, have difficult content, and are large in volume. While the curriculum has been painstakingly established, teachers seem hard-pressed to implement it. It is therefore no easy task to get them to understand and implement a student-focused style of teaching that encourages children to think and work problems out on their own.

The educational system in Egypt is changing, and in 2002, the elementary schools changed from a five-year to a six-year system. A sixth grade was introduced from September in 2004. Also, the content of the fourth and fifth grade curricula is to be spread out over fourth to sixth grades. Since this will create greater scheduling flexibility, it will create an environment in which it will be easier to introduce the kind of problem-solving oriented lessons we are envisioning. Also, the system of subject-based teacher assignments starts in the middle years (third and fourth grades) of elementary school, so it is common for teachers to teach mathematics not only at the local elementary school, but at the junior and senior high schools as well.

2. Teacher Training

There have been active efforts to provide training for mathematics teachers in recent years. Evidence to this effect includes efforts by the Egyptian government to change mathematics education, such as its request to Japan for assistance in this area. Egypt is urgently translating the GB that we created into Arabic and is considering to use it to train teachers. We have actually already created a digest version of the GB, and are now using it intensively to conduct training for mathematics teachers in 27 Egyptian provinces. The organization that served as our counterpart in the mini-project is playing a central role in this training. For three years, from 2002 to 2004, they are going to serve as lecturers and others involved in teacher training programs whose participants thus far have primarily been inspectors and senior teachers. The European Union and World Bank are providing support to enable Egyptian teachers to participate in this training. The training content covers the current state of mathematics education in Japan and the significance of problem solving and related curricular content, followed by specific exercises using actual materials. The teachers approach the sessions very enthusiastically.

3. Research lessons

In Egypt, there is virtually no Lesson Study between teachers to develop their lessons or investigate materials to use in their classes.

Last December, we invited counselors from the Egyptian Ministry of Education, inspectors, as well as teachers from experimental schools in our project to attend a Lesson Study at one of those experimental schools. The study attracted 50 teachers from the area. It would be not easy to say that the teachers gained an adequate understanding of problem-solving lessons, and the class teacher and students were rather nervous during the lesson since it was the first time for them to participate in a study lesson. Naturally, therefore, there were plenty of areas that will require additional attention. However, this occasion created a venue for teachers to discuss their lessons, and it inspired teachers at that school to work even more enthusiastically on their lessons afterwards. It was an extremely meaningful activity in terms of the future of this project.

In the discussions after the lesson, there were mainly questions and comments focused on the "pros" and "cons" of the inspector system and on the topic of problem solving.

Case 10: Lesson Study in Kenya
Efforts to Establish a Teacher Training System through the Strengthening of Mathematics and Science in Secondary Education (SMASSE) Project[1]

Takuya Baba

1. Problem Recognition at Start-Up

Although various problems were identified at the time this project was launched, a basic survey conducted in 1998 indicated that the key to solving the problems lay in finding ways to combat children's passive attitude toward mathematics and science. The typical class tended to be a teacher-led class in which students would be given a formula, asked to solve some sample questions and then left to spend extended periods of time solving practice problems. Significantly, this kind of class tends to cultivate a passive attitude toward science and mathematics among the students.

To better understand what was happening in classrooms before this project was launched, a basic survey was conducted in 1998 using questionnaires, interviews and classroom observations. In the interviews, teachers said that the key to a successful class was mathematical activities and the sharing of ideas between students, but they pointed to *students' passive attitudes* as preventing this from happening.

In the twelve classes observed, however, few strategies were observed that aimed to improve the students' passive attitudes were made by teachers through making students speak or discuss their ideas.

2. Purpose and Specific Features of Activities

In light of these problems, the goal of this project was to improve education by creating a system of training for those currently teaching. Nine out of 71 districts were selected to serve as pilot

---

[1] SMASSE enlarged to the SMASSE-WECSA in Africa.
See http://www.smasse.org/

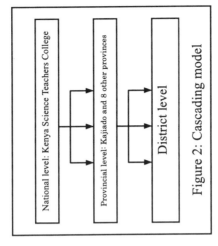

Figure 2: Cascading model

National level: Kenya Science Teachers College

Provincial level: Kajiado and 8 other provinces

District level

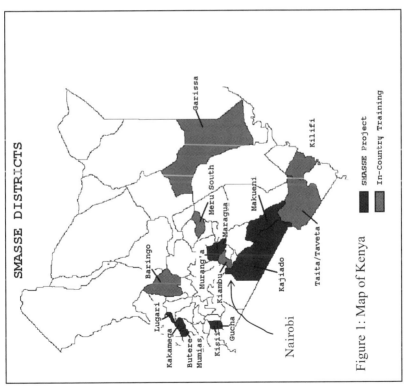

Figure 1: Map of Kenya

districts (Figure 1), and a cascading model was adopted to implement the project at multiple levels: the national level, provincial level, and district level (Figure 2). The project was designed to target about 2,000 science and mathematics teachers.

The training aimed to improve teaching knowledge and skills, to inspire teachers to help students take an active approach and to use these developments to change lessons and change the students' passive attitude. The essence of these goals is expressed in the ASEI principle (Activities, Students, Experiments and Improvisation). In mathematics, measures were taken to get children to look reflectively at their own environment and to think and express themselves based on their own experiences by incorporating an open-ended approach and by giving consideration to social and cultural factors.

3. Results

Our efforts since 1998 are beginning to show some results. The introduction of ownership and the sustained preservation of capital in Kenya have started to be seen. Budgeting is happening at the national level. At the provincial level, local government agencies and associations of school principals have begun working together to ensure that the limited resources available for education are being used effectively.

As a result of the joint activities between the Kenyans and the Japanese specialists, training materials have been created, and classes utilizing the ASEI principle (Photo 1) are slowly being implemented. It takes time to conduct mock classes and hold sessions to study teaching materials, but the teams that are doing this are achieving real improvements in teaching.

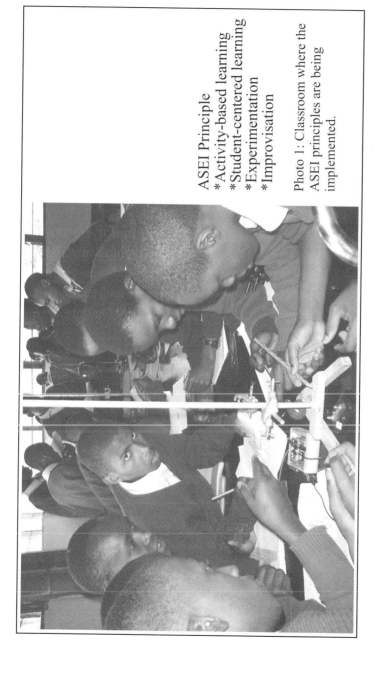

ASEI Principle
* Activity-based learning
* Student-centered learning
* Experimentation
* Improvisation

Photo 1: Classroom where the ASEI principles are being implemented.

Case 11: Lesson Study in Ghana

Minoru Yoshida

I visited Ghana in 1997 and 2000 and observed the actual mathematics lessons there. Since I stayed for only a short time during each visit, two weeks and three weeks, respectively, I was not able to observe very many lessons. However, I would like to describe several of the lessons I learned from observing those classes.

First, in spite of an enthusiastic desire to learn among the children, the lessons were not conducted in a correspondingly enthusiastic manner. Second, unlike Japanese lessons, lessons in Ghana were teacher-led and authoritative in nature. In contrast to the approach adopted by Japanese teachers, who smile perhaps more than is necessary, these teachers seemed instead to have an air of majesty about them. Third, I observed several similarities to Japanese classes in terms of the classroom culture. For example, when confirming an answer, teachers in Japan often ask, "Is this correct?" and the students respond, "Yes, that's correct." The response style was also evident in Ghana. That is, when confirming an answer, Ghanaian students would clap their hands all at once to express their collective agreement. The question of why such similar phenomena exist in both Japan and Ghana could be an important topic to examine in a study of classroom culture.

Based on these three phenomena, I would like to present several issues that will require further attention and research in order to improve the lessons in Ghana. First, under conditions in which students cannot study at home due to a shortage of textbooks and notebooks, they have to acquire all their academic skills during

Figure 1: GES - JICA STM Project at Shinshu University
GES: Ghana Education Service
STM: Science Technology and Mathematics

class time at school. It is therefore especially important that teachers clarify the lesson goals and academic skills that they are trying to impart to the students in each class hour.

Second, related to the first issue, it is important for teachers to practice basic teaching techniques, such as introducing topics, asking questions, and writing on a chalkboard. Third, and also significantly related to the first issue, investigations must be made on how Ghanaian teachers ascertain the student's readiness.

The baseline survey, completed in 2000, pointed in particular to the need to improve academic skills in the areas of "unit concepts," "concepts of ratios and proportions," and "the integration of the relationship between numbers and graphs," all of which are important for cultivating the "practical abilities" emphasized in Ghana. It is important that efforts be made to clarify whether teachers recognize this need and whether they are addressing it in their lessons.

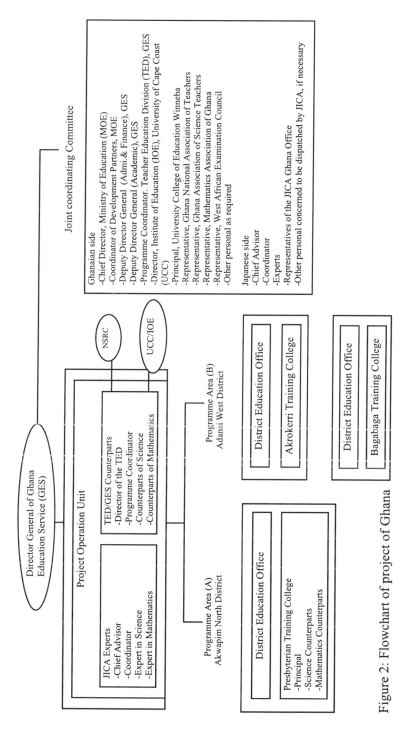

Figure 2: Flowchart of project of Ghana

Case 12: Lesson Study in South Africa
How is Lesson Study Promoted in South Africa?

Katsunori Hattori

Since 1999, Naruto University of Education, through a partnership with Hiroshima University, has participated in the Mpumalanga Secondary Science Initiative (MSSI) in South Africa, a project sponsored by JICA. The goals of this project are to improve the teaching skills of current teachers, reform classes by creating a system of in-service teacher training at the school level and to improve students' understanding of science and mathematics.

1. MSSI Overview

1.1 Implemented by
   Mpumalanga Department of Education

1.2 Supported by
   Japan International Cooperation Agency (JICA), the University of Pretoria, Hiroshima University, and Naruto University of Education

1.3 Target population
   Current mathematics and science teachers in secondary schools (540 schools) in Mpumalanga Province

1.4 Development Approach
   (a) Foster a sense of ownership/partnership, (b) emphasize experience-based models of teaching, (c) make improvements through evaluations (baseline surveys, monitoring evaluations).

1.5 Implementation
   (a) Mathematics teacher training (Hiroshima University, Naruto University of Education), (b) workshop development (Mpumalanga Province), (c) development of in-school training and creation of a lesson study framework, (d) creation and use of a study guide to improve classes.

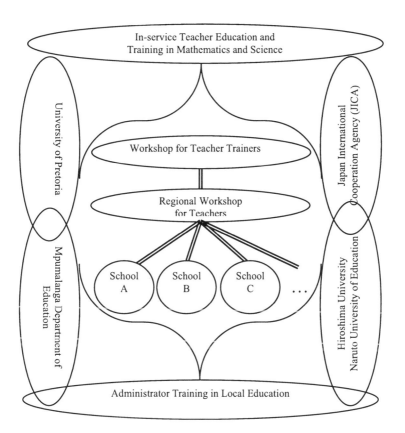

Figure 1: The MSSI system

2. Implementation of Lesson Study

2.1   Positioning of Lesson Study

MSSI is structured as shown in Figure 1. Lesson Study does not refer just to in-school training, but to a process by which mathematics teachers at several schools in the same community work together to research teaching materials, develop teaching plans (lesson plans) and practice study lessons. Participants in training programs organized by the local education authorities (school administrators) who had trained at Hiroshima University by JICA support these activities.

2.2   Guidebook Use

The guidebook (Figure 2), based on the training in Japan, carefully lays out approaches to Lesson Study that can be used for workshops, in-school training sessions, and within many school collaboration. It also describes the order in which components are to be undertaken in an easy-to-understand manner.

2.3   Revitalization of Lesson Study

It is gradually becoming common for study meetings to be held after lessons are developed, based on the approaches and methods described above. Vigorous discussions take place in a relaxed atmosphere.

Figure 2: Lesson Study guidebook for workshops

Photo 1: Mathematics Lesson Study meeting

Case 13: Lesson Study in Honduras
How can we improve the Teaching Skills of teachers who have missed out on a College Education?

Eiichi Kimura

Lesson development methods are proposed by using the National Textbooks and Guidebooks.

1. The Role of the JICA's PROMETAM Project[1]

Currently in Honduras, the basic qualification for elementary school teachers is being revised from what is called the teacher's level of school education (the equivalent of completion of high school in Japan) to require college education. A "continuous teacher training" (PFC) program is being conducted, offering college courses in various regions to enable teachers who are interested to obtain either a junior college or university degree after completing the course. The "Improvement in Technical Teaching in the Area of Mathematics Project" (PROMETAM) is a JICA technical cooperation project that serves as part of the training program. Since August 2002, JICA members have served as instructors in five locations in three prefectures (Figure 1), conducting courses on the curricula for first-, second- and third-year students (Photo 1 and Figure 2). The course materials used are the student workbooks (workbooks that accompany the textbooks) and the teacher's guidebook (which includes proposed lessons for every unit and every hour of class) of the Honduras National Textbook created under the leadership of JICA members, senior volunteers and education experts, based on Japanese teaching methods at the National Institute for Research and

---

[1] It enlarged to the Regional Project "I Love Mathematics!" in Central America and the Caribbean.

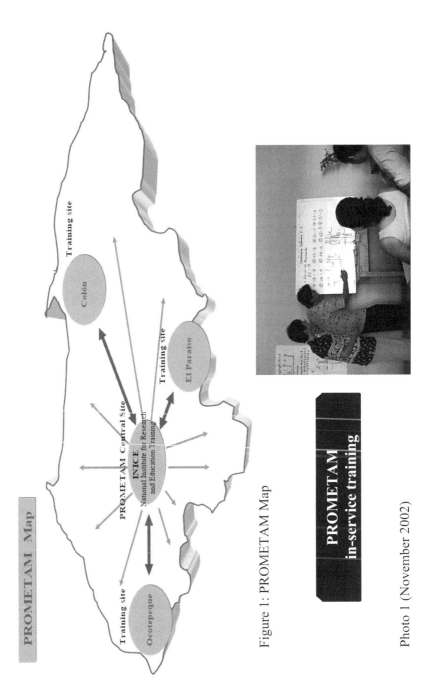

Figure 1: PROMETAM Map

Photo 1 (November 2002)

Education Training (INICE)[2]. These teaching materials will also be used in PFC programs conducted by Honduran teachers outside the PROMETAM project.

2. Goals of PROMETAM

PROMETAM aims to facilitate the development of "Classes in which students proactively engage in problem-solving activities" and the "number of student activity hours" is the index by which this development is gauged.

This methodology may be considered normal in developed countries where the provision of resources, such as each student having his or her own textbook, is taken for granted. In many countries, this is simply not the case. In these countries, there may be many other resources missing or in scarce supply in mathematics classrooms. If a school or parents cannot afford to buy textbooks for each student, the teacher has little choice but to write everything on the blackboard and have students copy. This is sometimes referred to as an "information-absorption" or "transmission" model of teaching.

Having workbooks for students as well as textbooks is an important component of PROMETAM. Only by making available these kinds of resources, including teachers' guidebooks, can we hope to change pedagogy.

Courses not only explain the mathematical and pedagogical background of the subjects and the use of teaching materials, but also incorporate lesson simulations so that teachers can learn how to develop their own lessons.

---

[2] It was established with support of JICA.

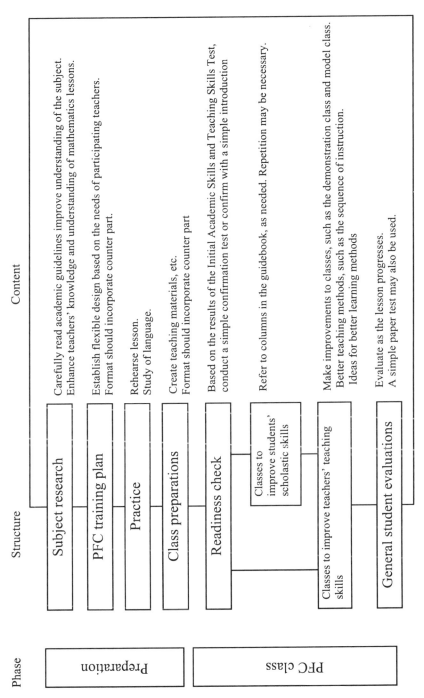

Figure 2: PFC Training Flow Chart (URL: http://www.prometam.hn2.com)

APPENDICES

Appendix 1: "Exploring the Unfolding of a Cylinder":

A 6th Grade Mathematics Lesson[1], by the Teacher Kozo Tsubota Attached Elementary School – University of Tsukuba

Abraham Arcavi

Summary of the lesson and its goals

This is the second of three lessons (assigned by the curriculum) on the unfolding of solids. During this class, the problem situation is to design models of unfolded cylinders and then to assemble them in order to check whether the models work (the folding indeed yields a cylinder). The goals are that students learn interactively (with concrete materials and with other students) about the structural components of the intervening two-dimensional figures, their relative positions, and, in certain cases, the importance of careful planning and measurements. In the process, students would exercise their imagination, spatial visualization abilities, and creativity.

The lesson develops as follows. First, the teacher reminds the class about the previous experience they had (during the first lesson on this topic) with the unfolding of a tetrahedron, and asks to think for a moment about the shape of the unfolding of a cylinder. After the class worked on the problem for a while, the teacher invites students to share their drawings on the blackboard. The first proposal is the classical: a rectangle and two tangent circles attached to its largest sides (prototypically, the largest sides are the horizontal). The teacher takes the opportunity to analyze the figure with the class, and to make sure students understand and agree on all the details. Thus, by asking several questions, simple but very important issues are raised and discussed, like:

---

[1] On the basis of an 11 minute video (of parts of the lesson) edited by the CRICED at the University of Tsukuba (http://www.criced.tsukuba.ac.jp/math/video/)

- the two circles (the bases of the cylinder) should be of the same size,
- the two circles should be tangent to the corresponding pair of parallel sides of the rectangle (and not secant to them),
- the length of the tangent sides should be equal to the circumference of the circles (students recall the number π and the formula for calculating the circumference),
- the length of the other two sides of the rectangle (to which the circles are not attached) are unconstrained (short sides and long sides will yield short or slender cylinders respectively),
- the points of tangency could be anywhere on each of the opposite sides of the rectangle.

Once these issues were discussed, the teacher encourages the class to produce alternative plane models for the unfolding of the cylinder. The first alternatives consist of cutting a piece from one side of the rectangle and reattaching it to the other side, in such a way that the area is conserved and the parts can match when the rectangle is folded to make the "body" of the cylinder. The class begins to propose other models including slicing and reattaching parts of the bases, and many other creative designs, many of which will not fold into a cylinder. At a certain point, the teacher encourages the students to actually cut their designs, attempt to fold them into a cylinder and see if they succeed. In case of failure students are encouraged to analyse the sources of their erroneous designs. By the end of the class, the models are displayed on the blackboard as a record of the activity.

Components of the lesson and main events in the class

The lesson has three distinct parts: the planning and designing phase, the practical work of assembling of the cylinders out of the models proposed, and the noticing and discussions of the failures.

An important feature of this class is that students are not provided with previous experiences cutting cylinders and observing the possible unfolding shapes. Only once in this lesson, the teacher shows how to cut a cylinder, as an illustration. Thus, students are forced to exercise their visualization skills in order to imagine the cut and the unfolding, or conversely, to imagine a shape that would fold into a cylinder. This mental activity is then contrasted with the actual practical work, in which students can revise their imagined figures. Some interesting mistakes arise, for example, figures that do not fold at all to a cylinder, or figures that would succeed provided their measures would have been taken into consideration.

An important teaching strategy was used by the teacher in the design of this class: having students make hypothesis, using their "mind's eyes", and then confirm or refute them, analysing back their plans. The commitment and motivation to one's hypothesis can be a fruitful source of learning, especially when these are refuted, provided a discussion is held to reflect about the whole process.

The teacher attached importance to students' display of creativity and to the sharing of ideas among the class.

Possible issues for discussion and reflection with teachers observing this lesson

As with any other lesson to be watched and reflected upon among teachers, a first issue would be to request teachers their opinions, and to start the dialogue from there. The following are some of the issues such a lesson may raise:

- *What may be the goals of this lesson?*
  This lesson may have several interwoven goals, and the development of the class could vary according to the emphasis one may put on each of them. Here are some goals one may want to discuss, prioritize, combine or discard:

development of spatial visualization, making/confirming/ revising hypothesis, stimulating creativity, developing manual/geometrical skills, developing a mathematical eye towards solids and boxes we see around and their properties, and having fun. Certainly other goals can be established.

- *How can we characterize the mathematics of this lesson?*
An interesting issue to discuss and reflect upon is the kind of mathematics and meta-mathematics students may learn in activities of this nature. This is not a lesson on Euclidean (deductive) geometry. Although there are some computational elements (for example, wrong measurements – resulting, in the circumference of the base being of different length than the side of the rectangle), these are not the main focus of the lesson either. So, what kind of geometry do students learn from such a lesson? As mentioned before, students may develop visualization skills and qualitative insights about relationships between elements that compose a whole. Also, as mentioned earlier, students may exercise hypothesis making, plan design, and executing practical work to confirm a hypothesis or to test the workability of their plans. One question to reflect upon would be the extent to which these kind of "qualitative" (as opposed to Euclidean or computational) geometry should be taught, how it may support other kinds of geometry learning, and how to take advantage of it in future lessons.

Appendix 2: "New Ways of Calculation":
A 3ʳᵈ Grade Mathematics Lesson[1], by the Teacher Yasuhiro Hosomizu Attached Elementary School – University of Tsukuba

Abraham Arcavi

The topic of this third grade lesson is around the calculation of a series of multiplications of two numbers between 20 and 30, in which their unit digits add up to 10 (e.g. 25x25; 24x26; 23x27; etc.). On the basis of such calculations, the goal of the lesson is to engage students in noticing patterns, finding an easier algorithm to perform the calculations, formulating a rule for such algorithm and explaining why it works. The new rule the students are supposed to find is: the result of such multiplications will be 600 plus the result of the multiplication of the two unit digits (e.g. 24 x 26 = 600 + 24). The rule can be easily formulated and justified in symbolic terms as follows:

$$(20 + a)(20 + (10 - a)) = 600 + a(10 - a)$$

Obviously, third grade students lack these tools. Therefore, in order to produce and/or understand an explanation, they will have to resort to numerical methods. The teacher, in his subsequent debriefing with colleagues (as part of the open forum of the Lesson Study), says that he is aware that third graders also lack the tools fourth graders have, such as the area model.

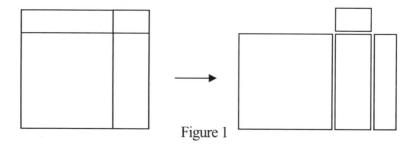

Figure 1

---

[1] On the basis of an 11 minute video (of parts of the lesson) edited by the CRICED at the University of Tsukuba (http://www.criced.tsukuba.ac.jp/math/video/)

(see Figure 1, above, in which the square represents 20x20, the two larger rectangles *together* represent 20x10 and the smaller rectangle represents the product of the units).

Therefore the teacher plans to discuss the explanation, in a subsequent lesson, by building with the students a discrete version of the area model, using beads.

Components of the lesson and main events in the class:

- Although the students are nudged to find regularities, the way in which they are randomly assigned the calculations to be performed (by drawing from a set of cards) does not make the observation of patterns a straightforward task.
- The first pattern some students notice is that the numbers to be multiplied add up to 50.
- After that, some students seemed to find the "rule" sought.
- A student disagreed with the generality of the rule by providing a counterexample: 19x31, whereas the factors add up to 50, the result is 589.
- With the help of the teacher, the class reformulated the original rule by restricting it to factors which add up to 50 but have a 2 as their tens digit.
- Towards the end of the class, a student explained why the hundreds digit of the result is always 6.

Possible issues for discussion and reflection with teachers observing this lesson

- *What may be the goals of this lesson?*
  By writing on the board the title of the activity (New ways of calculation), the teacher sets the goal for the activity which is to find new algorithms. This *declared* goal is only part of the teacher's *intended* goal, which is much broader but not stated explicitly to the students: to learn to observe and express regularities, to formulate new rules, to check if they work, to generalize or constrain according to examples and counterexamples, to formulate the generalization, and to try to explain at least in part why the rule works.

Therefore, the teacher must lead the class beyond the finding of the new way of calculating. For some students, the discovery may be in itself the culmination of the work, but for many others, the question "why" seems natural. Why did the teacher choose to announce the goal of the class in the way he did anyway? Probably because he wants to catch students' interest and engagement in finding "easier" ways to calculate or to discover nice "tricks", and not to overwhelm them with big statements of goals, whereas those are implicit and will be pursued anyway.

- *How can we characterize the mathematics of this lesson?*

As in many Japanese lessons, arithmetic skills are a central component. However, they never seem to constitute a goal in themselves, they are the backbone upon which to build explorations which in turn enhance students' numerical skills. As a result of this exploration they will have a new way of calculating certain multiplication and along the way they will have practiced in a thoughtful way the multiplication algorithm in many examples and towards a goal, not as a mere sequence of practicing problems. The "doing" of mathematics is worth noticing throughout this lesson and it includes: noticing patterns, formulating them, making a hypothesis, checking examples and counterexamples, and attempting explanations. This blending of procedural practice with doing significant (a non-trivial) mathematics seems to be at the core of many Japanese mathematics lessons.

- *How does the teacher view his students?*

The teacher is aware that he is relying on many ideas that will come up from the class, and he is also very aware of the tools they don't have (area models of multiplication) to make sense of an explanation. Nevertheless, he seems to proceed with confidence towards what he expects and he beliefs students may be led to produce an explanation even with their limited background.

- *What are the characteristics of the classroom management of this teacher?*

  Much has been said about the potential benefits of classroom discussions of mathematical ideas, however such discussions may present several dilemmas for the teacher. For example, the goal of the discussion is that students present their own ideas, and the class follows them up in order to understand others' ideas, to check their correctness, to present alternatives and to collectively produce knowledge. However, if the brightest students present advanced ideas very early, most of the other students may not be able to engage in a conversation with them. The teacher's concern is to be as inclusive as possible, and that may require special ways of handling the discussion by giving the right to speak to different students, by slowing the pace, by restating ideas, by requesting clarifications from students about other students' ideas, and by avoiding early value judgments about correctness. Managing such discussion requires skill and experience.

- *Is there more mathematics at stake in this problem, of which the teacher should be aware of?*

  There is no evidence that the teacher expected another mathematical outcome, for example that the result of a multiplication of two numbers which add up to 50, is 625 minus the square of the distance of the factors from 25. It is indeed farfetched to expect that third graders will notice that. Nevertheless, what about considering mathematical implications and ramifications of the mathematics underlying the task, even if those are very unlikely to emerge in the class?

- *What may be the learning outcomes and the follow-up for such a lesson?*

  From the video, it is difficult to know what students learned. It would be interesting to analyze the different kinds of follow up teachers may plan for such a lesson, including the possibility of not following it up directly and immediately.

Appendix 3: "I Understand What You Want to Say":
A 5th Grade Mathematics Lesson[1], by the Teacher Yasuhiro Hosomizu
Attached Elementary School – University of Tsukuba

Abraham Arcavi

The topic of this lesson is division of decimal numbers, and more specifically, how one operation of division can be derived (or connected) to other operations for which the result is known. The teacher starts the lesson by posing the general "open expression" ___ $\div 3 =$ and fills in the open space with 5.4. Initially, students propose three ways to solve this division problem:

-   multiply 5.4 by 10, to obtain 54, divide 54 by 3 to obtain 18, and then divide 18 by 10 to obtain 1.8;
-   multiply both 5.4 and 3 by 10, to obtain $54 \div 30$, which yields 1.8;
-   think of 5.4 as meters and transform it into 540 cm, divide it by 3 to obtain 180 cm, and then convert it back to meters.

The teacher takes care that all the students understand the methods, and devotes some time to proper ways of writing ("representing") the processes, which he calls "the rule of division". Then the teacher proposes to fill in the blank with 2.7, and to calculate $2.7 \div 3$. One student suggests that $27 \times 0.1 = 2.7$ and proposes to use that knowledge to solve the exercise, other students directly converted 2.7 into either 27 or 270, and adjusted correctly the result. The teacher asks the class if there were other solutions, and one student proposes that 5.4 is twice 2.7, and then

---

[1] On the basis of an 11 minute video (of parts of the lesson) edited by the CRICED at the University of Tsukuba (http://www.criced.tsukuba.ac.jp/math/video/)

rely on the previous result of 5.4 ÷ 3 for the solution sought. The teacher summarizes with the class the ways in which the knowledge of one result can be used to solve new exercises, and makes explicit the essential similarity between multiplying and dividing by 10 (5.4 ÷ 3 → 54 ÷ 3 = 18 → 5.4 ÷ 3 = 1.8) and multiplying and dividing by 2 (2.7 ÷ 3 → 5.4 ÷ 3 = 1.8 → 2.7 ÷ 3 = 0.9). The conclusion is that one may use a known exercise to create and solve new ones, and the teacher proposes the students to do so. Towards the end of the lesson, the teacher scans the students' proposals (15.12 – which is characterized by some students as a "miscalculation" –, 0.35, 10.8, 8.1, 3.24, 1.8) and promises to check them during the next lesson.

Possible issues for discussion and reflection with teachers observing this lesson

- *What may be the goals of this lesson?*
   At first glance, the obvious goal for this lesson is the practice of division with decimals. However, it is clear from the very beginning that, although time is devoted to calculations, this class is not only about the practice of an algorithm to obtain correct results. A main goal is to uncover, discuss and apply properties of division which shed light about the conceptual understanding involved, and at the same time to use this understanding in order to facilitate calculations by connecting different exercises via the properties explored. In other words, this class is an illustration of how procedural and conceptual proficiency in arithmetic can be integrated within one extended activity. Thus, a topic for this discussion with teachers

would be to analyze how the teacher weaves procedural and conceptual knowledge, and how the design of such a lesson can be applied to others. Another possible topic for discussion would be the possible ways of planning a follow up lesson on the basis of all the suggestions provided by the students at the end.

- *How can we characterize the mathematics of this lesson?*

    As stated above, the mathematics of this lesson integrates procedural and conceptual knowledge, doing and explaining, proposing alternative solutions and their rationale, generalizing properties of operations, and applying the generalizations to new situations.

- *How does the teacher view his students?*

    As in many other Japanese lessons, the respect the teacher has for his students is implicit in most of his actions. Firstly, students' ideas are always brought to the fore, explained, discussed more than once. As a way to enhance classroom communication, the teacher requests a student to explain an idea more than once. Many times he requests other students to explain the idea, to ensure that most of the students understand each other. Secondly, he allows for other solutions, and capitalize on them (for example the way the $2.7 \div 3$ was connected to $5.4 \div 3$, emerged from the class and allowed the teacher to state a generalization for the "rule of division"). Thirdly, when proposing an open problem, he scans the class for different answers, which he promises to be the basis for the next lesson. There are some indirect evidences that students feel comfortable with open discussions and with voicing their opinions. For example, one student criticized the conversion of $5.4 \div 3$ to $54 \div 30$, because it makes it more difficult, whereas the purpose (as he thought) was to simplify the calculation. Furthermore, when the teacher scans the solutions, he pauses for a little while on each of them, thinks about the numbers suggested, smiles and says that he understands what they say, or in other words that he can trace their thinking processes.

Appendix 4: "How Many Blocks?"
A 1st Grade Mathematics Lesson[1], by the Teacher Hiroshi Tanaka
Attached Elementary School – University of Tsukuba

Aida Yap

The topic of this first grade lesson is on determining the number of blocks in the pile wherein some of the blocks are hidden from one's view. The main objective of this lesson is to engage students in visualizing the number of blocks in the pile and explaining how they got their answers. In order to determine the number of blocks in the pile, the students have to rely on their visualization skills.

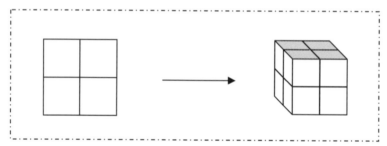

Components of the lesson and main events in the class

- A pile of blocks and a camera are hidden from the students. The front view of a pile of blocks is shown on the television screen. The teacher then asked the students to determine the number of blocks in the pile. This part of the lesson encourages the students to guess because showing the front view of the pile of blocks is quite deceiving. Most of the students answered 4 blocks, which was not surprising at all.
- The teacher afterwards positions the camera at a different angle so that another view of the pile of blocks is shown on

---

[1] On the basis of an 11 minute video (of parts of the lesson) edited by the CRICED at the University of Tsukuba (http://www.criced.tsukuba.ac.jp/math/video/)

the television screen. As before, the teacher asks the students to determine the number of blocks they think there are in the pile.

- A drawing of what was shown on the television screen was posted on the board. The teacher distributed a worksheet to each student. The worksheet contained the same drawing as the one posted on the board. The students were then asked to write their formulation and answer in the worksheet.

- Students came up with different mathematical formulations such as $4 + 4$, $3 + 2 + 3$, $1 + 3 + 4$, $4 + 3 + 1$, and $2 + 2 + 2 + 2$. Some students were asked to explain their work in front of the class.

- Towards the end of the lesson, the teacher brought out 8 big blocks and arranged them in a pile similar to what was shown in the drawing. The students were asked to come closer to the front so that they could clearly see the pile of blocks. The teacher repeated the explanation of some students using these blocks.

Possible issues for discussion and reflection with teachers observing this lesson:

- *What may be the goals of this lesson?*

  By showing the front view of the pile of blocks and writing on the board the question that students need to answer (How many blocks are there in the pile?) the teacher sets the goal of the activity. The teacher is not actually interested on whether the students come up with the correct number of blocks in the pile but rather on the students' way of thinking in getting the number of blocks in the pile.

- *How can we characterize the mathematics of this lesson?*

Visualization skill is a very important skill that any student must possess. Thus, giving problems that help develop the visualization skill of the students is really important even at this very early stage in elementary mathematics. Encouraging students to explain or defend his/her answer is really more important than the answer itself. In this way, the teacher would be able to discover student's mathematical thinking and possible misconceptions that the student may have. Corrections on the erroneous ways of thinking of the student may then be made accordingly.

- *How does the teacher view his students?*

The teacher is challenging the students all the time to imagine the number of cubes in the pile. Never did the teacher say that the answer given by the student is correct or not. It is evident that the teacher is not after the answers given to him by the students but rather on the thinking or reasoning behind each answer. The teacher feels confident that even at this early age students would be able to show evidence of their visualization skill.

- *What are the characteristics of the classroom management of this teacher?*

The teacher made use of a combination of strategies to get students attention all the time. He writes, explains, poses problems/questions, and process students' answers. The use of television to enhance his instruction was really a good idea to challenge the students to think. During the lesson proper, the teacher showed expertise in handling the discussion. After a student presented his/her work, the teacher always followed-up student's explanation.

It is very evident that the teacher was able to capture

students' attention through the activities he presented. Students really enjoyed the hands-on and minds-on activities given to them by the teacher. There was never a lull period during the discussion.

- *Is there more mathematics at stake in this problem of which the teacher should be aware of?*

 The teacher obviously attained his intended goal for this lesson. It would be interesting to find out the reasoning behind the other mathematical formulations made by the students. It is farfetched to expect these first graders to come up with mathematical formulations involving multiplication.

- *What may be the learning outcomes and the follow-up for such lesson?*

 From the video, it is clear that the students were able to come up with different ways of counting the number of blocks in the pile. In all these mathematical formulations, the visualization skill of the students is being challenged. It would be very worthy of note to find out if students can deal with counting the number of blocks in the pile containing more than 10 blocks or when there are more blocks hidden from the students view.